COLD SHOCK RESPONSE AND ADAPTATION

JMMB Symposium Series Volume 2

The reviews published in this book were originally presented as a written symposium in the Journal of Molecular Microbiology and Biotechnology Vol. 1 No. 2 (November 1999)

Edited by:

Masayori Inouye

and

Kunitoshi Yamanaka

Robert Wood Johnson Medical School
Department of Biochemistry
675 Hoes Lane
Piscataway
NJ 08854, USA

Copyright © 2000
Horizon Scientific Press
32 Hewitts Lane
Wymondham
Norfolk NR18 0JA
England

www.horizonpress.com

British Library Cataloguing-in-Publication Data

A catalogue record for this book is available from the British Library

ISBN: 1-898486-24-7

All rights reserved. No part of this publication may be reproduced, stored in a retrieval system, or transmitted, in any form or by any means, electronic, mechanical, photocopying, recording or otherwise, without the prior permission of the publisher. No claim to original U.S. Government works.

Printed and bound in Great Britain
by IBT Global Linited, London.

Contents

Contents		iii
1.	Introduction *M. Inouye*	1
2.	Cold Shock Response in *Escherichia coli* *K. Yamanaka*	5
3.	Cold Shock Response in *Bacillus subtilis* *P. L. Graumann, and M. A. Marahiel*	27
4.	Cold Acclimation and Cold Shock in Psychrotrophic Bacteria *M. Hebraud, and P. Potier*	41
5.	Responses to Cold Shock in Cyanobacteria *D. A. Los, and N. Murata*	61
6.	Molecular Responses of Plants to Cold Shock and Cold Acclimation *C. Guy*	85
7.	Cold Shock Response in Mammalian Cells *J. Fujita*	113
Index		143

Books of Related Interest

Gene Cloning and Analysis: Current Innovations   1997
Brian C. Schaefer (Ed.)

An Introduction to Molecular Biology   1997
Robert C. Tait

Genetic Engineering with PCR   1998
Robert M. Horton and Robert C. Tait (Eds.)

Prions: Molecular and Cellular Biology   1999
David A. Harris (Ed.)

Probiotics: A Critical Review   1999
Gerald W. Tannock (Ed.)

Peptide Nucleic Acids: Protocols and Applications   1999
Peter E. Nielsen and Michael Egholm (Eds.)

Intracellular Ribozyme Applications: Principles and Protocols   1999
John J. Rossi and Larry Couture (Eds.)

NMR in Microbiology: Theory and Applications   2000
Jean-Noël Barbotin and Jean-Charles Portais (Eds.)

Molecular Marine Microbiology   2000
Douglas H. Bartlett (Ed.)

Oral Bacterial Ecology: The Molecular Basis   2000
Howard K. Kuramitsu and Richard P. Ellen (Eds.)

Prokaryotic Nitrogen Fixation: A Model System   2000
Eric W. Triplett (Ed.)

For further information on these books contact:
Horizon Scientific Press, P.O. Box 1, Wymondham, Norfolk, NR18 0EH, U.K.
Tel: +44(0)1953-601106. Fax: +44(0)1953-603068. Email: mail@horizonpress.com

Our Web site has details of all our books including full chapter abstracts, book reviews, and ordering information:
www.horizonpress.com

1

Introduction

Masayori Inouye

Department of Biochemistry
Robert Wood Johnson Medical School
675 Hoes Lane, Piscataway,
New Jersey 08854-5635, USA

The most common stress that living organisms constantly confront in nature is likely to result from temperature changes. Figure 1 shows the temperature changes in Newark, New Jersey during a one year period. The graph shows both average highest and lowest temperatures of each month as indicated by open and solid circles, respectively. One can see that per month there is a difference of approximately 10°C between the highest and the lowest temperatures and annually the temperatures fluctuate from −5 to 30°C. How do organisms living in this area respond and adapt to the wide temperature changes, in particular, the low-temperature stress or the cold shock? It is interesting to note that these organisms are not exposed to high-temperature

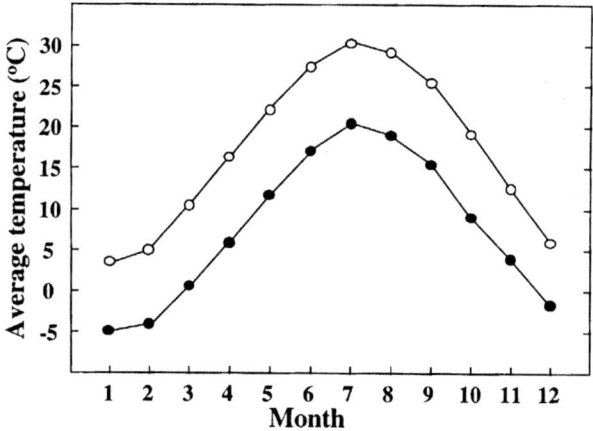

Figure 1. Annual Temperature Changes in Newark, New Jersey. Average highest and lowest temperatures for each month are shown by open and solid circles, respectively. Information was obtained from: http://weather.yahoo.com/almanac/Newark-NJ-US.

stress (heat shock). Nevertheless, while heat-shock effects have been extensively studied in both prokaryotes and eukaryotes, few studies have been carried out to show how living organisms respond to cold shock.

A major reason why heat shock is more thoroughly investigated as compared to cold shock is that heat shock causes a well-defined damage to cells, *i.e.*, protein unfolding or denaturation and that all living organisms from bacteria to humans contain heat-shock proteins, which are induced upon heat shock to assist protein folding. In contrast, cold shock does not produce such well-defined damage. As temperature is downshifted, cell growth slows down and eventually stops.

The research on cold shock raises a number of questions such as which cellular function is affected most upon cold shock, what makes cell growth stop, and whether there are well-conserved or common cold-shock proteins as in case of heat-shock proteins. These questions are not less important compared to those in case of heat shock, and for the last several years interesting results on cellular adaptation and response to cold shock have been obtained.

It is highly important to arrange a symposium of cold-shock response and adaptation to overview the present status of the cold-shock research. I believe that this is the first symposium organized on cold-shock research that encompasses bacteria to humans. In this symposium, Dr. K. Yamanaka, Dr. P. Graumann and Dr. M. A. Marahiel present the molecular mechanisms of cold-shock response and adaptation in *Escherichia coli* and *Bacillus subtilis,* the most studied systems at present. In these systems, it has been shown that there are a number of cold-shock proteins that are induced upon temperature downshift. In particular, small molecular weight proteins consisting of approximately 70 amino acid residues are produced in a large amount; CspA for *E. coli* and CspB for *B. subtilis*, which are proposed to function as an RNA chaperone to enhance translation by blocking the formation of secondary structures in mRNAs. We also asked Drs. M. Hebraud and P. Potier to present current studies on psychrotrophic bacteria, which preferentially live at low temperatures. Their research deals with an interesting question of how the optimal temperature range for growth is determined, and what determines the lowest permissive temperature for these bacteria.

The research on psychorotrophic and psychrophilic bacteria is also important for human health because they cause food spoilage and cause food-borne diseases.

Drs. D. A. Los and N. Murata present the research on desaturases of cyanobacteria, which play an essential role to maintain the membrane fluidity upon cold shock. Further, Dr. C. Guy presents an overview on the cold-shock research in plants, describing that there are a large number of cold-shock genes identified in plants, and a signal transduction pathway triggered by cold shock. He also discusses important questions such as how to improve plants to be more tolerant to low temperatures. Finally, Dr. J. Fujita, who found a cold-shock inducible RNA-binding protein called CIRP in mammalian cells, presents an exten-sive review on cold-shock response in mammalian cells.

Although each system seems to respond to cold shock in its unique way, studies from different organisms revealed existence of a few common cellular functions, which have to be dealt with for cold-shock adaptation. One such function is to increase the membrane fluidity and the other is to enhance translation probably at the level of initiation and elongation. I hope that the audience of the present symposium recognizes cold-shock research as a hot topic and that its further advancement will not only yield new exciting discoveries with respect to biological function, but will also aid the improvement of human health and the energy conservation.

We owe the success of this symposium to Dr. K. Yamanaka, who took a major editorial role in this event.

2

Cold Shock Response in *Escherichia coli*

Kunitoshi Yamanaka

Department of Biochemistry
Robert Wood Johnson Medical School
675 Hoes Lane, Piscataway, NJ 08854, USA

Abstract

Sensing a sudden change of the growth temperature, all living organisms produce heat shock proteins or cold shock proteins to adapt to a given temperature. In a heat shock response, the heat shock sigma factor plays a major role in the induction of heat shock proteins including molecular chaperones and proteases, which are well-conserved from bacteria to human. In contrast, no such a sigma factor has been identified for the cold shock response. Instead, RNAs and RNA-binding proteins play a major role in cold shock response. This review describes what happens in the cell upon cold shock, how *E. coli* responds to cold shock, how the expression of cold shock proteins is regulated, and what their functions are.

Introduction

All living organisms have developed sophisticated strategies to respond to several environmental stresses, such as osmolarity, pH, nutrition deprivation and thermal stresses. In terms of thermal stress, heat shock stress (temperature upshift) has been extensively studied and a common strategy how to respond to it has been emerged from bacteria to human (Hendrick and Hartl, 1993; Gottesman *et al.*, 1997). In contrast, cold shock stress (temperature downshift) was poorly understood, particularly in higher eukaryotes. To know the strategy how to respond to cold shock stress includes not only biological interests but also economic and health implications. Prevention of the spoilage of foods kept in a refrigerator, which is usually caused by contaminated bacteria, is clearly an important issue in our life.

Cold shock stress causes bacterial cells two major facts; a decrease in membrane fluidity and translational block. The former fact can be overcome by increasing an unsaturated fatty acid, consequently increasing a diunsaturated phospholipid in membrane. A specific set of proteins called cold shock proteins are transiently induced to overcome the translational block. Unlike heat shock proteins, no common cold shock proteins except for CspA-like proteins were identified among bacteria. Most of the free living bacteria possess at least one cold-shock-inducible CspA-like protein, whose function was proposed to be an RNA chaperone (Yamanaka et al., 1998; Graumann and Marahiel, 1998). Cyanobacteria and even human were revealed to contain cold-shock-inducible RNA-binding proteins, although they are another type from CspA (Sato and Nakamura, 1998; Nishiyama et al., 1997). RNA-binding proteins and RNAs thus play a central role in cell growth under low temperature conditions.

Cold Shock Response of Membrane Lipid Composition

Membranes are normally in a liquid crystalline form and undergo a reversible transition to a gel phase upon temperature downshift. To compensate for the transition from a fluid state to a nonfluid state, many organisms have developed mechanisms to change the membrane lipid composition, that is, fatty acid composition, by means of one or a combination of the following changes: an increase in fatty acid unsaturation, a decrease in average chain length, an increase in methyl branching, and an increase in the ratio of *anteiso*-branching relative to *iso*-branching.

It was first reported that *E. coli* adjusts its fatty acid composition in response to a lower growth temperature by increasing the amount of *cis*-vaccenic acid and decreasing the amount of palmitic acid incorporated into membrane phospholipid (Marr and Ingraham, 1962). ß-ketoacyl-ACP synthase II, encoded by *fabF*, plays a key role in the change of fatty acid composition and is responsible for elongating palmitoleic acid to *cis*-vaccenic acid (Garwin et al., 1980). Thus, upon cold shock, C16:1Δ9 is converted to C18:1Δ11, giving an increase in unsaturated fatty acids, and consequently, diunsaturated phospholipids which lowers the melting point and has a greater degree of flexibility comparing to saturated phospholipids. This type of response is named as homeoviscous adaptation (Sinensky, 1974). It is interesting to notice that the synthesis of ß-ketoacyl-ACP synthase II is not induced upon cold shock, but the enzyme activity is induced at low temperature (Garwin and Cronan Jr., 1980; Garwin et al., 1980). Mutants lacking this enzyme are deficient in both *cis*-vaccenic acid synthesis and thermal regulation (Garwin et al., 1980), indicating that the changes in membrane lipid composition are critical for bacterial adaptation to low temperatures.

In contrast, *B. subtilis* contains a membrane-bound fatty acid desaturase, encoded by *des*, which desaturates palmitate to *cis*-Δ5-hexadeceonate (Aguilar et al., 1998). The *des* expression is regulated at the level of transcription upon cold shock (Aguilar et al., 1998). The *des* deletion mutant grew well at both high and low temperatures, but showed severely reduced survival during stationary phase (Aguilar et al., 1998). In a cyanobacterium

Synechococcus sp. PCC7002, expression of *des* genes, *desA*, *desB* and *desC*, is regulated by a combination of mRNA synthesis and stabilization upon cold shock (Sakamoto and Bryant, 1997).

Cold Shock Proteins of *E. coli*

When an *E. coli* culture at 37°C is transferred to 10 or 15°C, a set of proteins called cold shock proteins are transiently induced (Jones *et al.*, 1987). These cold shock proteins are conventionally classified into two groups based on their expression patterns (Thieringer *et al.*, 1998): Class I cold shock proteins are expressed at an extremely low level at 37°C but are dramatically induced upon cold shock. Class I includes CspA, CspB, CspG, CspI, CsdA, RbfA, NusA, and PNP; ClassII cold shock proteins are expressed at a certain extent at 37°C and are induced moderately upon cold shock. ClassII includes IF-2, H-NS, RecA, α subunit of DNA gyrase, Hsc66, HscB, trigger factor, dihydrolipoamide acetyltransferase and pyruvate dehydrogenase (lipoamide).

Several cold shock proteins have been detected not only in *E. coli* but also in many other bacteria, including mesophilic bacteria, *B. subtilis* (Lottering and Streips, 1995; Graumann *et al.*, 1996), *Enterococcus faecalis* (Panoff *et al.*, 1997), and *Lactococcus lactis* (Panoff *et al.*, 1994), psychrotrophic bacteria, *Pseudomonas fragi* (Hebraud *et al.*, 1994; Michel *et al.*, 1996; 1997), *Arthrobacter globiformis* (Berger *et al.*, 1996), *Bacillus psychrophilus* (Whyte and Inniss, 1992), *Vibrio vulnificus* (McGovern and Oliver, 1995), *Pseudomonas putida* (Gumley and Inniss, 1996) *Listeria monocytogenes* (Phan-Thanh and Gormon, 1995; Bayles *et al.*, 1996), and *Rhizobium* (Cloutier *et al.*, 1992), and psychrophilic bacterium *Aquaspirillum arcticum* (Roberts and Inniss, 1992). It has yet to be determined how well cold shock proteins are conserved among bacteria.

Energy Generation

Two cold shock proteins, dihydrolipoamide acetyltransferase and pyruvate dehydrogenase (lipoamide), are subunits of pyruvate dehydrogenase complex along with dihydrolipoamide reductase (NAD^+). Pyruvate dehydrogenase complex is essential to convert pyruvate to acetyl CoA, which is an initial, important substrate for the TCA cycle as well as fatty acid synthesis. This suggests that the metabolic pathway to generate energy, ATP, is regulated upon cold shock.

Chromosome Dynamics

Negative supercoiling of plasmid DNA in *E. coli* cells has been shown to transiently increase upon cold shock (Goldstein and Drlica, 1984: Mizushima *et al*, 1997). DNA gyrase and HU protein play an important role in this DNA supercoiling reaction. This notion is consistent with that DNA gyrase α subunit is a cold-shock inducible protein (Jones *et al.*, 1992b). Thus, the regulation of chromosomal DNA supercoiling upon cold shock may be important to maintain DNA transactions such as replication, transcription and recombination. It is interesting to note that RecA, which is involved in recombination and repair, and H-NS, which is a nucleoid-associated DNA binding protein and is involved in gene expression and chromosome compaction, are also cold-shock inducible proteins (Jones et al., 1987). Growth inhibition at low temperature was observed in E. coli strains carrying hns mutations (Dersch et al, 1994). Expression of hns and gyrA has been demonstrated to be regulated by CspA in such a way as CspA may help or stabilize an open complex formation for transcription (La Teana et al., 1991; Jones et al., 1992b; Brandi et al., 1994). It is likely that changes in DNA superhelicity upon cold shock might be involved in the induction of cold shock response.

Protein Molecular Chaperones at Low Temperature

Previously protein folding or refolding is not considered to be a major issue at low temperatures. However, it becomes the important issue at present. Hsc66 and HscB, a DnaK and DnaJ homologue, respectively, are specifically induced upon cold shock (Lelivelt and Kawula, 1995). Synthesis of several proteins were affected by an hsc66 mutation, although cell growth at low temperature was not significantly impaired (Lelivelt and Kawula, 1995). Another cold-shock protein, trigger factor (TF), is a molecular chaperone functioning as a peptidyl-prolyl isomerase (Kandror and Goldberg, 1997; Scholz et al., 1997). It associates with nasent polypeptides on ribosomes, binds to GroEL, enhances GroEL's affinity for unfolded proteins, and promotes degradation of certain polypeptides (Kandror et al., 1995; 1997; Stoller et al., 1995; 1996). TF has been shown to be important for viability at low temperatures. When E. coli cells are stored at 4°C, they lose viability at an exponential rate. Cells with reduced TF content die faster, while cells overexpressing TF showed greater viability (Kandror and Goldberg, 1997). These results suggest that the proper protein folding and the refolding of cold-shock-damaged proteins are as important upon cold shock as they are upon heat shock. It should be noticed that as similar to TF, a peptidyl-prolyl cis-trans isomerase of B. subtilis, encoded by ppiB, is cold-shock inducible (Herrler et al., 1994; Graumann et al., 1996) and is found to be involved in protein folding (Göthel et al., 1998).

Adaptation of Translation Factory upon Cold Shock

At low temperature, protein synthesis, especially the translation initiation step, becomes rate limiting for cell growth, resulting in accumulation of 70S ribosomes (Friedman et al., 1971; Broeze et al., 1978; Farewell and Neidhardt, 1998). The addition of several translation inhibitors such as chloramphenicol has been shown to result in the induction of cold-shock proteins and the addition of other translation inhibitors such as kanamycin induces heat-shock proteins (VanBogelen and Neidhardt, 1990). Thus, it has been proposed that the ribosome is a physiological sensor for thermal stresses, both cold shock and heat shock (VanBogelen and Neidhardt, 1990). The level of (p)ppGpp has also been shown to be involved in the cold-shock response, which reduces the level of (p)ppGpp (Mackow and Chang, 1983; Jones et al., 1992a).

Upon cold shock, there is a growth lag period called the acclimation phase before cell growth resumes (Jones et al., 1987). During the acclimation phase, most of the cellular protein synthesis is blocked most probably at the translation initiation step. However, the synthesis of cold-shock protein is able to bypass this translational block during this period. The synthesis of non-cold-shock proteins require some translational factors that are induced upon cold shock (Jones and Inouye, 1996). These factors include IF-2, which is a translation initiation factor that lets the initiation tRNA (fMet-tRNA) bind to the 30S subunit (Jones et al., 1987), CsdA, which is a DEAD-box protein of helicaces and associates with ribosomes (Jones et al., 1996), and RbfA, which is a free 30S ribosome binding factor required for optimal growth particularly at low temperature (Jones and Inouye, 1996). The cold-shock ribosome adaptation model has been proposed (Jones and Inouye, 1996). In this model, at high temperatures ribosomes are translatable for all cellular mRNAs. Upon cold shock, translation initiation is transiently blocked, resulting in a decrease in polysomes and an increase in 70S, 50S and 30S ribosomes, which are incapable for translation. Cold-shock proteins whose function is related to translation are synthesized to convert cold-sensitive, non-translatable ribosomes to cold-adapted, translatable ribosomes, resulting in the recovery of the cellular protein synthesis and cell growth. As previously proposed that the ribosomes function as the sensor for thermal stresses, the translation is a key factor for adaptation to a given low temperature.

How, then, is the expression of cold-shock genes able to be induced during the acclimation phase upon cold shock, while the synthesis of most of non-cold-shock proteins is blocked? An important feature of the cold shock induction of CspA is the presence of the downstream box (DB) sequence in its mRNA (Mitta et al., 1997; Etchegaray and Inouye, 1999b) (Figure 1). The DB sequence was originally proposed to serve as an independent translational signal besides the Shine-Dalgarno (SD) sequence and is located downstream of the initiation codon in the coding sequence of mRNAs (Sprengart et al., 1990; 1996). The DB sequence is complementary to the region called the anti-DB sequence of the 16S rRNA. It is speculated that formation of a duplex between the DB sequence of mRNA and the anti-DB sequence of 16S rRNA is responsible for translational enhancement (Sprengart et al., 1996). When the DB sequence was deleted from the *cspA* gene, CspA induction upon cold shock was not observed (Mitta et al., 1997). Furthermore, when the DB sequence was inserted into the ß-galactosidase gene, it became cold-shock inducible (Mitta et al., 1997; Etchegaray and Inouye, 1999b). These results clearly indicate that the DB sequence plays a crucial role in the cold shock induction. It is worth mentioning that not only *cspA* but also other cold-shock genes belonging to the Class I group, whose expression is dramatically induced upon cold shock, contain the DB sequence downstream of the initiation codon in their mRNAs (Mitta et al., 1997). The

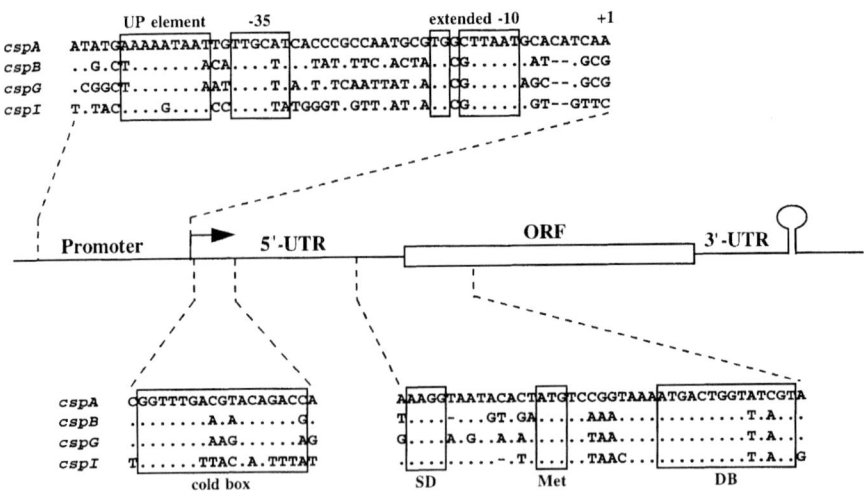

Figure 1. Sequence Comparison of *csp* Genes. Comparison of characteristic regions of *cspA*, *cspB*, *cspG*, and *cspI* genes. Characteristic motifs are boxed. SD, Met and DB represent Shine-Dalgarno sequence, translation initiation codon, and downstream box, respectively. Nucleotides identical to *cspA* are shown as dots and gaps are shown as dashes. At the middle, schematic structure of these genes is presented. See text in details.

formation of translation initiation complex between mRNA for Class I cold-shock genes and 16S rRNA is likely enhanced by the presence of the DB sequence in concert with the SD sequence (Etchegaray and Inouye, 1999b). The DB sequence is therefore essential for the induction of cold-shock proteins during the acclimation phase, when the translation initiation for non-cold-shock proteins is blocked. In other words, mRNAs for Class I cold-shock genes are efficiently translated without any requirement of cold-shock proteins, whose function is related to translation capacity, such as IF-2, CsdA and RbfA.

It is interesting to note that when a translatable truncated *cspA* mRNA is overexpressed at low temperature, cell growth is completely blocked (Jiang et al., 1996b). This effect was termed LACE (low-temperature antibiotic effect of truncated *cspA* expression). It is considered to be caused by the truncated *cspA* mRNA, which still possesses the DB sequence. This mRNA may efficiently form an initiation complex, in such a way as almost all ribosomes are trapped by the overproduced truncated *cspA* mRNAs. In fact, under LACE, a high polysome profile, which is normally reduced upon cold shock, was retained, resulting in production of only truncated CspA products (Jiang et al., 1996b). This also supports the notion that the DB sequence is crucial for efficient translation upon cold shock.

In contrast, it is not well understood how the Class II cold-shock genes are induced. It should be noticed that *hns* and *gyrA* are regulated by CspA at the level of transcription upon cold shock, that is, CspA may help or stabilize the open complex formation for transcription. It is possible that the changes in DNA superhelicity may be involved in cold-shock induction of these genes.

Regulation of *cspA* Expression

The *cspA* gene is located at 80.1 min on the *E. coli* chromosome and it is monocistronically transcribed in a clockwise direction. The most important feature is that the *cspA* mRNA possesses an unusually long 5'-untranslated region (5'-UTR) consisting of 159 bases (Tanabe *et al.*, 1992). Since the *cspA* mRNA as well as CspA protein are hardly detected at 37°C, it was suggested that *cspA* expression is regulated at the level of transcription (Tanabe *et al.*, 1992). Since then, *cspA* expression has been extensively analyzed, and it is currently considered to be regulated at the levels of transcription, mRNA stability and translation.

The *cspA* gene has a strong promoter equipped with the UP element as well as the extended -10 sequence, which is active at both 37°C and 15°C (Fang *et al.*, 1997; Goldenberg *et al.*, 1997; Mitta *et al.*, 1997) (Figure 1). When the *cspA* promoter was replaced with the *lpp* promoter, a constitutive, strong promoter of a major outer membrane protein, *cspA* expression was still cold-shock inducible, indicating that the cold-shock induction of CspA does not depend on its promoter (Fang *et al.*, 1997). However, the expression level at 15°C is higher with the *cspA* promoter than that with the *lpp* promoter (Fang *et al.*, 1997), suggesting that the *cspA* promoter is one of the strongest promoter in *E. coli*. The *cspA* promoter is equipped with two unique motifs: one is the presence of an AT rich sequence immediately upstream of the -35 region (Goldenberg *et al.*, 1997; Mitta *et al.*, 1997). The AT rich sequence, called the UP element, was reported to be directly recognized by the α subunit of RNA polymerase and to confer a strong transcription activity (Ross *et al.*, 1993). In fact, when the UP element was deleted from the *cspA* promoter, the promoter activity was almost diminished (Mitta *et al.*, 1997). The second unique motif is the presence of a TGn motif immediately upstream of the -10 region. It was reported that this motif together with the -10 region constitute a so called extended -10 region and if a promoter contains the extended -10 region, the -35 region becomes dispensable (Kumar *et al.*, 1993). Moreover, the *cspA* promoter itself was shown to be active even at 37°C, at which the CspA protein is hardly detected (Mitta *et al.*, 1997). Thus, by virtue of these two motifs, the *cspA* promoter has a strong activity both at 37°C and 15°C. It should be mentioned that transcription of the *cspA* gene does not require any *de novo* protein synthesis upon cold shock (Etchegaray and Inouye, 1999a). Note that unlike the heat-shock response, no specific sigma factor responsible for cold shock has been identified among prokaryotes.

It is very important to notice that the *cspA* mRNA has an unusually long 5'-UTR (Tanabe *et al.*, 1992) and is extremely unstable at 37°C (half life; <12 sec) (Brandi *et al.*, 1996; Goldenberg *et al.*, 1996; Fang *et al.*, 1997). Immediately upon cold shock, the *cspA* mRNA becomes stable (half life; >20 min) (Brandi *et al.*, 1996; Goldenberg *et al.*, 1996; Fang *et al.*, 1997). Again, this stabilization of the *cspA* mRNA upon cold shock does not require any *de novo* protein synthesis (Etchegaray and Inouye, 1999a). In the 5'-UTR, there is a putative RNaseE cleavage site immediately upstream of the SD sequence. This site is considered to be responsible for the extreme instability of the *cspA* mRNA, since the three-base substitution mutation at this region resulted in dramatic stabilization of the mRNA, allowing a high CspA production even at 37°C (Fang et al., 1997). When the region from +26 to +143 of the 5'-UTR was deleted from the translational *cspA-lacZ* fusion construct, high ß-galactosidase activity was obtained even at 37°C (Mitta et al., 1997). Thus, it is clear that the mRNA stabilization plays a major role in *cspA* expression upon cold shock and that the unusually long 5'-UTR is responsible for its instability at 37°C, in other words, the 5'-UTR makes the *cspA* mRNA extremely unstable at 37°C. As described above the *cspA* mRNA is constitutively transcribed even at 37°C but can not be translated due to its extreme instability. However, the exact mechanism of the *cspA* mRNA stabilization upon cold shock is not known. RNase, which is responsible for the degradation of the *cspA* mRNA, may be somehow inactivated upon cold shock. Alternatively, the secondary structure of the *cspA* mRNA at 37°C may be highly susceptible to RNase but not be accesible to ribosomes, while the secondary structure of the *cspA* mRNA at low temperature may be changed to be accesible to ribosomes. It is worth mentioning that although the *cspA* mRNA becomes stable and is accumulated at the nonpermissive temperaturte in the temperature-sensitive RNaseE mutant, CspA production was not detected under this condition (Fang et al., 1997), suggesting that in addition to the stability of mRNA, the 5'-UTR may have another role. This may be related to the secondary structure of the *cspA* mRNA, its translation efficiency and the transcription attenuation.

The 5'-UTR of the *cspA* mRNA contains a unique sequence called the cold box, which may form a stable stem-loop structure (Jiang et al., 1996a) (Figure 1). CspA is transiently produced during acclimation phase upon cold shock. However, when the 5'-UTR of the *cspA* mRNA or the fragment containing the cold box sequence was overexpressed at 15°C, CspA production became no more transient. When CspA was simultaneously overproduced, however, the normal transient expression was resumed (Jiang et al., 1996a). Moreover, when the *cspA* gene without the cold box sequence was reintroduced into a *cspA* deletion mutant, CspA production was again poorly repressed at the end of the acclimation phase (Bae et al., 1997; Fang et al., 1998). These results clearly indicate that the cold-box sequence plays an important role in autoregulation of *cspA* expression.

Very recently, it has been reported that CspA is also transiently induced during early exponential phase after dilution of overnight culture at 37°C to a level of approximately 100,000 molecules/cell (Brandi et al., 1999). Therefore, they claimed that the designation of CspA as a major cold-shock protein is a misnomer. However, upon cold shock CspA is induced at a level of 800,000 to 1,000,000 molecules/cell, 8 to 10 times higher than the transient expression at 37°C (Jiang et al., 1997). Furthermore, when the temperature is lowered less than 10°C, CspA and its homologues CspB and CspG are major proteins induced with little production of all the other cellular proteins (Etchegaray et al., 1996). It is important to note that well-established heat-shock proteins such as DnaK and GroE are known to be constitutively expressed under normal growth conditions (Yura et al., 1993). In addition, heat-shock proteins can be induced without heat shock if other stresses such as alcohol are given. Similarly, it has been shown that CspA can also be induced at 37°C in the presence of chloramphenicol (VanBogelen and Neidhardt 1990; Jiang et al., 1993; Etchegaray and Inouye 1999a). Clearly CspA can be termed as a cold-shock protein, and it is the major cold-shock protein in E. coli.

Structure and Function of CspA

CspA was originally identified as a major cold-shock protein in E. coli and consists of 70 amino acid residues (Goldstein et al., 1990). It has been purified and its three-dimensional structure has also been determined by both X-ray crystallography (Schindelin et al., 1994) and NMR spectroscopy (Newkirk et al., 1994; Feng et al., 1998). It exists as a monomer and consists of five anti-parallel ß-strands, ß1 (K5-N13), ß2 (F18-T22), ß3 (D29-H33), ß4 (Q49-E56) and ß5 (A63-L70), forming a ß-barrel structure with two ß-sheets (ß1-ß2-ß3 and ß4-ß5) (Schindelin et al., 1994; Newkirk et al., 1994; Feng et al., 1998). Five of the hydrophobic residues (V9, I21, V30, V32, and V51) form the hydrophobic core in the ß-barrel structure (Feng et al., 1998). It is particularly important to mention that the ß-sheet consisting of ß1 to ß3 contains seven out of eight of aromatic residues (W11, F12, F18, F20, F31, F34 and Y42) and two Lys residues (K10 and K16). Furthermore, on this surface there are two RNA-binding motifs, RNP1 ($K_{16}GFGFI_{21}$) on ß2 and RNP2 ($V_{30}FVHF_{34}$) on ß3. CspA has been shown to bind to both single-stranded DNA (ssDNA) and RNA (Jiang et al., 1997). Hydrophilic interactions between CspA and nucleic acid occur through aromatic residues on this surface of the CspA molecule. In fact, it has been demonstrated that mutations of Phe residues on this surface severly affected the DNA-binding activity (Hillier et al., 1998). In the case of B. subtilis, CspB, a major cold shock protein, has been shown to be highly homologous to E. coli CspA (Willimsky et al., 1992). Its three-dimensional structure has also been resolved to form a similar ß-barrel structure (Schindelin et al., 1993; Schnuchel et al., 1993). Mutations of Phe residues in the RNA-binding motifs has been shown to abolish the DNA-binding activity (Schröder et al., 1995). These results clearly indicate that the ß-sheet of ß1 to ß3, especially RNA-binding motifs, plays a crucial role in the binding of CspA to nucleic acids.

E. coli CspA exists as a monomer (Schindelin et al., 1994; Newkirk et al., 1994), while B. subtilis CspB exists as a dimer, which is formed by intermolecular hydrogen bonds between ß4 strands (Schindelin et al., 1993; Schnuchel et al., 1993). What could be the determinant for the dimer formation? It is speculated that the ß4 strand in E. coli CspA is somehow masked by its N-terminal region, because E. coli CspA has an extra 3 amino acid residues at its N-terminal region comparing to that of B. subtilis CspB (Newkirk et al., 1994). Since a mutant CspA, which is lacking the N-terminal 4 amino acid residues, still exists as a monomer, it is unlikely that the N-terminal region is involved in the inhibition of dimer formation (Wang et al., manuscript in preparation). Alternatively, the structural differences, which are observed in the ß4 strand and the loop connecting strand ß4 and ß5, probably preclude dimer formation in *E. coli* CspA (Makhatadze and Marahiel, 1994). Interestingly, CspD, which is a member of the *E. coli* CspA family but is not cold-shock inducible as discussed below, was found to exist as a dimer (Yamanaka and Inouye, unpublished). By analyzing several chimeric proteins between CspA and CspD, it was recently found that only the β4 strand is enough to determine the dimer formation. When the β4 strand of CspA was replaced with the β4 strand of CspD, the resulting chimeric CspA became to be a dimer (Yamanaka and Inouye, unpublished). Whether a critical amino acid residue in the β4 strand exists or the structure of the β4 strand is required for dimerization remains to be addressed.

CspA binds cooperatively to RNA and ssDNA without an apparent sequence specificity (Jiang *et al.*, 1997). Binding of CspA to RNA renders it more sensitive to RNase. It is thus proposed that CspA functions as an RNA chaperone to facilitate translation at low temperatures by preventing the formation of stable secondary structures in mRNAs (Jiang *et al.*, 1997). In fact, in the cell free translation system, which was prepared from cells grown at 37°C, translation of the *cspA* mRNA was shown to be enhanced by the addition of the purified CspA (Brandi et al., 1996). Note that the CspA binding to RNA is rather weak (Jiang *et al.*, 1997), so that ribosome movement on mRNA would not be hampered by the CspA binding.

The CspA Family of *E. coli*

The entire genome sequence of *E. coli* has been determined. It possesses totally nine CspA homologues, CspA to CspI (Yamanaka *et al.*, 1998). The lengths vary from 69 to 74 amino acid residues and the identities between each two proteins vary from 29 to 83%. The secondary structure analysis by using the Chou-Fassman's method predicts that all of them likely form a β-barrel structure as shown for CspA (Yamanaka *et al.*, 1998). Moreover, the RNA-binding motifs, RNP1 and RNP2, are also well conserved among them except for CspF and CspH, which have S16, K18, Q31, and V34 or I34 instead of K16, F18, F31 and F34 for CspA, respectively, suggesting that the *E. coli* CspA homologues are likely to bind to nucleic acid (Yamanaka *et al.*, 1998).

Among nine homologues, only four, CspA (Goldstein *et al.*, 1990), CspB (Lee *et al.*, 1994; Etchegaray *et al.*, 1996), CspG (Nakashima *et al.*, 1996), and CspI (Wang *et al.*, 1999), are cold-shock inducible. Based on the similarities of their primary amino acid sequences, they may have a similar structure and function as CspA does. It should be noted that although the *cspA* gene has been shown to be dispensable at both high and low temperatures, the increased production of CspB and CspG was observed upon cold shock in the *cspA* deletion mutant, suggesting that the CspA function may be at least partially complemented by CspB and CspG (Bae *et al.*, 1997). Thus, CspA, CspB, CspG, and CspI may have overlapping functions each other, if not the same.

In terms of regulation of gene expression, *cspA*, *cspB*, *cspG*, and *cspI* genes were revealed to share several important features (Figure 1). (i) They all contain the UP element upstream of the -35 region as well as the extended -10 region in their promoter sequences, which are considered to maintain the high promoter activity even at low temperatures (Goldenberg et al., 1997; Mitta et al., 1997; Wang et al., 1999). (ii) They all contain an unusually long 5'-UTR in their mRNAs (159, 161, 156, and 145 bases for cspA, cspB, cspG, and cspI, respectively) (Tanabe et al., 1992; Etchegaray et al., 1996; Nakashima et al., 1996; Wang et al., 1999). As shown in the case of cspA, it is believed that these 5'-UTRs play a crucial role in their cold-shock inducibility. (iii) They all contain the cold box at the 5'-end region of their 5'-UTRs, which plays a role in autoregulation to repress their own gene expression at the end of the acclimation phase (Jiang et al., 1996a; Bae et al., 1997; Fang et al., 1998; Wang et al., 1999). (iv) They all contain a DB sequence downstream of the translation initiation codon, which plays an essential role in the cold-shock induction by enhancing translation (Mitta et al., 1997; Etchegaray and Inouye, 1999b). Taken altogether, it is likely that expression of all four csp genes is regulated essentially in the same manner. It should be mentioned, however, that the optimal temperature ranges for the induction of these genes are different (Etchegaray et al., 1996; Wang et al., 1999). CspA induction occurs over the broadest temperature range (30°C to 10°C), CspI induction occurs over the narrowest and lowest range (15°C to 10°C), and CspB and CspG occurs at temperatures between the above extremes (20°C to 10°C). The nucleotide sequences of their 5'-UTR are different, suggesting differences in their mRNA secondary structure and therefore in their stabilities. The 5'-UTR of the cold-shock inducible csp genes exert a negative effect on their expression at 37°C, affecting their mRNA stabilities and translation initiation efficiencies (Mitta et al., 1997). These are likely to be differently modulated upon cold shock, depending upon the secondary structure of each mRNA. A smaller temperature difference woud be enough for cspA expression, while a larger temperature difference might be required for cspI expression.

CspC and CspE were originally isolated as multicopy suppressors for a temperature-sensitive chromosome partition mutant and are expressed at 37°C (Yamanaka et al., 1994). CspE was also shown to be involved in chromosome condensation (Hu et al., 1996) and to inhibit the transcription antitermination mediated by Q protein of phage λ (Hanna and Liu, 1998). In a cspE deletion strain, the synthesis of a number of proteins at 37°C was found to be altered comparing to the wild-type strain. Interestingly, cspA expression was derepressed at 37°C (Bae et al., 1999). The derepression of cspA in the cspE mutant occured at the level of transcription in a promoter-independent manner but was not caused by stabilization of the cspA mRNA, which is a major cause of CspA induction upon cold shock (Bae et al., 1999). In vitro transcription assays demonstrated that CspE may increase the efficiency of transcription pausing through direct binding to either the ssDNA or the nascent RNA of the cold box region of cspA (Bae et al., 1999). Note that CspE can bind to ssDNA and RNA (Bae et al., 1999). Moreover, CspE has been shown to be able to interact with elongation complexes containing RNAs as short as 10 nucleotides through protein-RNA interactions (Hanna and Liu, 1998). These results indicate that CspE functions as a negative regulator for cspA expression at 37°C.

cspD expression is not cold-shock inducible but is induced during the stationary phase and upon carbon starvation (Yamanaka and Inouye, 1997). It was also shown to be inversely dependent on growth rate. *cspD* expression was suggested to be regulated at the level of transcription rather than translation. However, it is independent of the stationary sigma factor σ^s (Yamanaka and Inouye, 1997). No significant defect was observed in the *cspD* deletion mutant and purified CspD can bind to ssDNA and RNA but not to dsDNA (Yamanaka and Inouye, unpublished). Interestingly, CspD exists as a dimer and the ß4 strand is found to be a determinant for dimer formation as described above. Cellular function of CspD is yet to be determined. Nothing is known about CspF and CspH.

The chromosomal location of nine genes for the *E. coli* CspA family are as follows: *cspE*, 14.2 min (c); *cspD*, 19.9 min (cc); *cspH*, 22.6 min (cc); *cspG*, 22.6 min (c); *cspI*, 35.3 min (cc); *cspB*, 35.3 min (cc); *cspF*, 35.3 min (c); *cspC*, 41.1 min (cc); and *cspA*, 80.1 min (c), where in parentheses c and cc mean that transcription is occured in clockwise and counter-clockwise direction, respectively (Rudd, 1998; Yamanaka *et al.*, 1998). The *cspA* gene located at 80.1 min is unique among the family by being closer to the DNA replication origin (84.6 min), while the others are clustered from 14.2 min to 41.1 min, where the DNA replication termination sites are also clustered. Based on these facts together with sequence similarities, it was proposed that the large CspA family of *E. coli* resulted from several steps of gene duplications (Yamanaka *et al.*, 1998). Why did *E. coli* duplicate *csp* genes several times? As mentioned above, the region where eight *csp* genes are located corresponds to the region where the DNA replication termination sites are clustered. The DNA replication termination has to be tightly controlled. This would be the reason the DNA replication termination sites are duplicated several times. If *csp* genes and the DNA replication termination sites are somehow linked, it is possible that these might be co-duplicated. It is interesting to mention that the clusters of *cspE-cspH-cspG* and *cspC-cspF-cspB* are located in a mirror image centered around 28.8 min, where TerA, one of the DNA replication termination site, is located (Yamanaka *et al.*, 1998).

Evolution of the Cold-Shock Domain

Proteins homologous to CspA are found widely in prokaryotes and to date, more than 90 CspA-like proteins have been identified including Gram-positive and Gram-negative bacteria, and psychrophilic, psychrotrophic, mesophilic and thermophilic bacteria. However, *csp*-like genes have not been found in the archaeal genomes (*Archaeoglobus fulgidus, Methanobacterium thermoautotrophicum, Methanococcus jannaschii,* and *Pyrococcus horikoshii*) and most of the parasites (*Borrelia burgdorferi, Chlamydia trachomati, Mycoplasma pneumoniae, Mycoplasma genitalium,* and *Treponema pallidum*), whose entire genomes have been sequenced. It can be considered that CspA-like proteins are not required in the parasitic bacteria, since host cells would possess their own response systems against any stresses. One exception is *Rickettsia prowazekii*, which has one *csp*-like gene (Andersson *et al.*, 1998). In addition, cyanobacteria also do not have a *csp*-like gene. However, it has another cold-inducible RNA-binding protein RbpA1, which does not belong to the CspA family but is similar to the eukaryotic RNA-binding proteins, such as U1A snRNP (Sato and Nakamura, 1998; Graumann and Marahiel, 1998). Interestingly, its expression is regulated by several DNA-binding proteins, which bind to the 5'-UTR, at the level of transcription.

Furthermore, eukaryotic Y-box proteins, such as human YB-1 and *Xenopus* FRGY-2, contain a region called cold-shock domain (CSD) with more than 40% identity to *E. coli* CspA (Matsumoto and Wolffe, 1998; Yamanaka *et al.*, 1998). These facts indicate that the CSD is well-conserved from prokaryotes to eukaryotes. Cellular functions of the Y-box protein family are diverged and include positive or negative transcriptional regulation (Kashanchi *et al.*, 1994; Ohmori *et al.*, 1996; Ting *et al.*, 1994; Zou *et al.*, 1995), translational masking of mRNAs (Ranjan *et al.*, 1993), signal transduction pathway (Duh *et al.*, 1995; Shinozaki and Yamaguchi-Shinozaki, 1996), cell proliferation (Landomery and Sommerville, 1995; Bargou *et al.*, 1997), and development (Moss *et al.*, 1997). It should be mentioned that proteins of Y-box protein family are relatively large comparing to those of CspA family by having N- and C-terminal extra domains, which are thought to be responsible for their diverged functions (Matsumoto and Wolffe, 1998; Graumann and Marahiel, 1998).

From the evolutionary aspect, it is interesting to note that the archaea, which may be an ancestor for eukaryotic nucleus, does not contain a *csp*-like gene, and that eukaryotes have proteins containing a CSD as mentioned above. Recently, a parasite *Rickettsia prowazekii*, which is considered to be more closely related to eukaryotic mitochondria than is any other microbe studied so far, was revealed to contain a *csp*-like gene (Andersson *et al.*, 1998). Thus, it can be speculated that the ancestor of the eukaryotic CSD might come from the ancestor of mitochondria, which may be a bacterial parasite. Later on, the gene for CSD was transferred from mitochondria to nucleus and then diverged, probably by duplication, gene arrangement, insertion and so on. It is also interesting to mention that *Mycobacterium tuberculosis* and *Mycobacterium leprae* possess two *csp*-like genes. One seems to encode a bacterial CspA homologue, while another seems to encode a eukaryotic Y-box protein homologue, since it contains a C-terminal extra domain, which possesses a similar feature as the C-terminal extra domain of human YB-1 protein. It consists of alternating regions of predominantly basic and acidic amino acid residues, which is proposed to function as a charge-zipper to mediate protein-protein interactions. Therefore, the mycobacterial homologue is interesting from aspects of both biological function and evolution of CSD.

Concluding Remarks

All living organisms have developed the mechanisms to respond to environmental stresses, such as temperature fluctuation. In the case of temperature upshift (heat shock response), the heat-shock sigma factor plays a crucial role in induction of heat-shock proteins (Yura et al., 1993). Heat-shock proteins, in particularly molecular chaperones and proteases, are well-conserved from bacteria to human (Hendrick and Hartl, 1993; Gottesman et al., 1997). This is consistent with the notion that protein folding and refolding and/or degradation are the most important issues upon heat shock. In contrast, in the case of temperature downshift (cold-shock response), no such a sigma factor has been identified. Synthesis of cold-shock proteins seems to be regulated mainly at the post-transcriptional level as described above. Thus, the fate of individual mRNA for each cold-shock protein plays a central role in cold-shock response. Most of the free living bacteria possess at least one cold-shock-inducible CspA homologue, which is likely function as an RNA chaperone (Jiang et al., 1997). Cyanobacteria contain another type of cold-shock-inducible RNA-binding protein rather than a CspA-like protein (Sato and Nakamura, 1998; Graumann and Marahiel, 1998). Furthermore, a cold-inducible RNA-binding protein has been identified in human cells (Nishiyama et al., 1997). RNA-binding proteins probably with an RNA chaperone activity play a major role in cell proliferation at low temperatures. In fact, it has been demonstrated that at least one out of three CspA homologues is required for cell growth in B. subtilis (Graumann et al., 1997). In the case of E. coli, there are nine CspA homologues and it is considered that their functions may overlap. Simultaneous introduction of mutations of nine genes into one cell seems to be a somewhat challenging task. However, a quadruple deletion mutant (cspA cspB cspE cspG) was obtained and did show a cold-sensitive growth (Xia et al., unpublished). These results indicate that RNA-binding proteins are essential for bacterial growth. These facts are consistent with the notion that translation block and stable secondary structure formation of DNA and RNA are the major issues upon cold shock.

It is interesting to notice that upon cold shock, expression of heat-shock proteins is repressed (Taura et al., 1989). When the heat-shock proteins are artificially produced at low temperatures, cellular survival was lost much faster (Kandror and Goldberg, 1997). These results suggest that cold-shock proteins and heat-shock proteins provide protection against opposite thermal extremes, but these two protectors are tightly controlled not to be existing simultaneously.

Very recently, it has been demonstrated that the rpoH mRNA secondary structure at 42°C, which is accessible to ribosomes, is different from that at 30°C, and that the rpoH mRNA structure itself is determinant for rpoH expression without any other factors (Morita et al., 1999). Thus, it was proposed that the rpoH mRNA acts as an RNA thermosensor (Morita et al., 1999). It was also suggested that lcrF mRNA of Yersinia pestis and the λ phage cIII mRNA might be other examples for RNA thermosensor (Storz, 1999). CspA is not produced at 37°C because of extreme instability of its mRNA, even though it is transcribed. However, upon cold shock the cspA mRNA becomes stable and translatable with no requirement of any de novo protein synthesis. Thus, the cspA mRNA is designed to be unstable and non-translatable at high temperatures, and simultaneously it is designed to be stabilized and translatable at low temperatures without any *de novo* protein synthesis. In this regard, mRNAs for cold-shock proteins, at least CspA homologues, can be considered to act as an RNA thermosensor, but in a different manner from above mentioned examples.

Acknowledgements

The author greatly thank Prof. Masayori Inouye for providing great opportunity to write this review, continuous warm support, and stimulating discussions. The author also thank Dr. S Phadtare for critical reading of this manuscript. This work was supported by a grant from the National Institutes of Health (GM19043) to Dr. M. Inouye.

References

Aguilar, P.S., Cronan, J.E., and de Mendoza, D. 1998. A *Bacillus subtilis* gene induced by cold shock encodes a membrane phospholipid desaturase. J. Bacteriol. 180: 2194-2200.

Andersson, S.G.E., Zomorodipour, A., Andersson, J.O., Sicheritz-Ponten, T., Alsmark, U.C.M., Podowski, R.M., Naslund, A.K., Eriksson, A.-S., Winkler, H.H., and Kurland, C.G. 1998. The genome sequence of *Rickettsia prowazekii* and the origin of mitochondria. Nature. 396: 133-140.

Bae, W., Jones, P.G., and Inouye, M. 1997. CspA, the major cold shock protein of *Escherichia coli*, negatively regulates its own gene expression. J. Bacteriol. 179: 7081-7088.

Bae, W., Phadtare, S., Severinov, K., and Inouye, M. 1999. Characterization of *Escherichia coli cspE*, whose product negatively regulates transcription of *cspA*, the gene for the major cold shock protein. Mol. Microbiol. 31: 1429-1441.

Bargou, R.C., Jurchott, K., Wagener, C., Bergmann, S., Metzner, S., Bommert, K., Mapara, M.Y., Winzer, K.J., Dietel, M., Dorken, B., and Royer, H.D. 1997. Nuclear localization and increased levels of transcription factor YB-1 in primary human breast cancers are associated with intrinsic MDR1 gene expression. Nature Med. 3: 447-450.

Bayles, D.O., Annous, B.A., and Wilkinson, B.J. 1996. Cold stress proteins induced in *Listeria monocytogenes* in response to temperature downshock and growth at low temperatures. Appl. Environ. Microbiol. 62: 1116-1119.

Berger, F., Morellet, N., Menu, F., and Potier, P. 1996. Cold shock and cold acclimation proteins in the psychrotrophic bacterium *Arthrobacter globiformis* SI55. J. Bacteriol. 178: 2999-3007.

Brandi, A., Pietroni, P., Gualerzi, C.O., and Pon, C.L. 1996. Post-transcriptional regulation of CspA expression in *Escherichia coli*. Mol. Microbiol. 19: 231-240.

Brandi, A., Pon, C.L., and Gualerzi, C.O. 1994. Interaction of the main cold shock protein CS7.4 (CspA) of *Escherichia coli* with the promoter region of *hns*. Biochimie. 76: 1090-1098.

Brandi, A., Spurio, R., Gualerzi, C.O., and Pon, C.L. 1999. Massive presence of the *Escherichia coli* 'major cold-shock protein' CspA under non-stress conditions. EMBO J. 18: 1653-1659.

Broeze, R.J., Solomon, C.J., and Pope, D.H. 1978. Effects of low temperature on *in vivo* and *in vitro* protein synthesis in *Escherichia coli* and *Pseudomonas fluorescens*. J. Bacteriol. 134: 861-874.

Cloutier, J., Prévost, D., Nadeau, P., and Antoun, H. 1992. Heat and cold shock protein synthesis in arctic and temperate strains of Rhizobia. Appl. Environ. Microbiol. 58: 2846-2853.

Dersch, P., Kneip, S., and Bremer, E. 1994. The nucleoid-associated DNA-binding protein H-NS is required for the efficient adaptation of *Escherichia coli* K-12 to a cold environment. Mol. Gen. Genet. 245: 255-259.

Duh, J.L., Zhu, H.A., Shertzer, H.G., Nebert, D.W., and Puga, A. 1995. The Y-box motif mediates redox dependent transcriptional activation in mouse cells. J. Biol. Chem. 270: 30499-30507.

Etchegaray, J.P., and Inouye, M. 1999a. CspA, CspB, and CspG, major cold shock proteins of *Escherichia coli*, are induced at low temperature under conditions that completely block protein synthesis. J. Bacteriol. 181: 1827-1830.

Etchegaray, J.P., and Inouye, M. 1999b. Translational enhancement by an element downstream of the initiation codon in *Escherichia coli*. J. Biol. Chem. 274: 10079-10085.

Etchegaray, J.P., Jones, P.G., and Inouye, M. 1996. Differential thermoregulation of two highly homologous cold-shock genes, *cspA* and *cspB*, of *Escherichia coli*. Genes Cells. 1: 171-178.

Fang, L., Hou, Y., and Inouye, M. 1998. Role of the cold-box region in the 5' untranslated region of the *cspA* mRNA in its transient expression at low temperature in *Escherichia coli*. J. Bacteriol. 180: 90-95.

Fang, L., Jiang, W., Bae, W., and Inouye, M. 1997. Promoter-independent cold-shock induction of *cspA* and its derepression at 37°C by mRNA stabilization. Mol. Microbiol. 23: 355-364.

Farewell, A., and Neidhardt, F.C. 1998. Effect of temperature on *in vivo* protein synthetic capacity in *Escherichia coli*. J. Bacteriol. 180: 4704-4710.

Feng, W., Tejero, R., Zimmerman, D.E., Inouye, M., and Montelione, G.T. 1998. Solution NMR structure and backbone dynamics of the major cold-shock protein (CspA) from *Escherichia coli*: evidence for conformational dynamics in the single-stranded RNA-binding site. Biochem. 37: 10881-10896.

Friedman, H., Lu, P., and Rich, A. 1971. Temperature control of initiation of protein synthesis in *Escherichia coli*. J. Mol. Biol. 61: 105-121.

Garwin, J.L., and Cronan J.E.Jr. 1980. Thermal modulation of fatty acid synthesis in *Escherichia coli* does not involve *de novo* enzyme synthesis. J. Bacteriol. 141: 1457-1459.

Garwin, J.L., Klages, A.L., and Cronan, J.E.Jr. 1980. ß-Ketoacyl-acyl carrier protein synthase II of *Escherichia coli*.: evidence for function in the thermal regulation of fatty acid synthesis. J. Biol. Chem. 255: 3263-3265.

Goldenberg, D., Azar, I., and Oppenheim, A.B. 1996. Differential mRNA stability of the *cspA* gene in the cold-shock response of *Escherichia coli*. Mol. Microbiol. 19: 241-248.

Goldenberg, D., Azar, I., Oppenheim, A.B., Brandi, A., Pon, C.L., and Gualerzi, C.O. 1997. Role of *Escherichia coli cspA* promoter sequences and adaptation of translational apparatus in the cold shock response. Mol. Gen. Genet. 256: 282-290.

Goldstein, E., and Drlica, K. 1984. Regulation of bacterial DNA supercoiling: plasmid linking numbers vary with growth temperature. Proc. Natl. Acad. Sci. USA. 81: 4046-4050.

Goldstein, J., Pollitt, N.S., and Inouye, M. 1990. Major cold shock proteins of *Escherichia coli*. Proc. Natl. Acad. Sci. USA. 87: 283-287.

Göthel, S.F., Scholz, C., Schmid, F.X., and Marahiel, M.A. 1998. Cyclophilin and trigger factor from *Bacillus subtilis* catalyze in vitro protein folding and are necessary for viability under starvation conditions. Biochem. 37: 13392-13399.

Gottesman, S., Wickner, S., and Maurizi, M.R. 1997. Protein quality control: triangle by chaperones and proteases. Genes Dev. 11: 815-823.

Graumann, P., and Marahiel, M.A. 1998. A superfamily of proteins that contain the cold-shock domain. Trends Biochem. Sci. 23: 286-290.

Graumann, P., Schröder, K., Schmid, R., and Marahiel, M.A. 1996. Cold shock stress-induced proteins in *Bacillus subtilis*. J. Bacteriol. 178: 4611-4619.

Graumann, P., Wendrich, T.M., Weber, M.H.W., Schröder, K., and Marahiel, M.A. 1997. A family of cold shock proteins in *Bacillus subtilis* is essential for cellular growth and for efficient protein synthesis at optimal and low temperatures. Mol.Microbiol. 25: 741-756.

Gumley, A.W., and Inniss, W.E. 1996. Cold shock proteins and cold acclimation proteins in the psychrotrophic bacterium *Pseudomonas putida* Q5 and its transconjugant. Can. J. Microbiol. 42: 798-803.

Hanna, M.H., and Liu, K. 1998. Nascent RNA in transcription complexes interacts with CspE, a small protein in *E. coli* implicated in chromatin condensation. J. Mol. Biol. 282: 227-239.

Hebraud, M., Dubois, E., Potier, P., and Labadie, L. 1994. Effect of growth temperatures on the protein levels in a psychrotrophic bacterium, *Pseudomonas fragi*. J. Bacteriol. 176: 4017-4024.

Hendrick, J.P., and Hartl, F.-U. 1993. Molecular chaperone functions of heat-shock proteins. Annu. Rev. Biochem. 62: 349-384.

Herrler, M., Bang, H., and Marahiel, M.A. 1994. Cloning and characterization of *ppiB*, a *Bacillus subtilis* gene which encodes a cyclosporin A-sensitive peptidyl-prolyl *cis-trans* isomerase. Mol. Microbiol. 11: 1073-1083.

Hillier, B.J., Rodriguez, H.M., and Gregoret, L.M. 1998. Coupling protein stability and protein function in *Escherichia coli* CspA. Folding Design. 3: 87-93.

Hu, K.H., Liu, E., Dean, K., Gingras, M., DeGraff, W., and Trun, N.J. 1996. Overproduction of three genes leads to camphor resistance and chromosome condensation in *Escherichia coli*. Genetics. 143: 1521-1532.

Jiang, W., Fang, L., and Inouye, M. 1996a. The role of 5'-end untranslated region of the mRNA for CspA, the major cold-shock protein of *Escherichia coli*, in cold-shock adaptation. J. Bacteriol. 178: 4919-4925.

Jiang, W., Fang, L., and Inouye, M. 1996b. Complete growth inhibition of *Escherichia coli* by ribosome trapping with truncated *cspA* mRNA at low temperature. Genes Cells. 1: 965-976.

Jiang, W., Jones, P., and Inouye, M. 1993. Chloramphenicol induces the transcription of the major cold shock gene of *Escherichia coli, cspA*. J. Bacteriol. 175: 5824-5828.

Jiang, W., Hou, Y., and Inouye, M. 1997. CspA, the major cold-shock protein of *Escherichia coli*, is an RNA chaperone. J. Biol. Chem. 272: 196-202.

Jones, P.G., Cashel, M., Glaser, G., and Neidhardt, F.C. 1992a. Function of a relaxed-like state following temperature downshifts in *Escherichia coli*. J. Bacteriol. 174: 3903-3914.

Jones, P.G., and Inouye, M. 1994. The cold-shock response - a hot topic. Mol. Microbiol. 11: 811-818.

Jones, P.G., and Inouye, M. 1996. RbfA, 30S ribosomal binding factor, is a cold-shock protein whose absence triggers the cold-shock response. Mol. Microbiol. 21: 1207-1218.

Jones, P.G., Krah, R., Tafuri, S.R., and Wolffe, A.P. 1992b. DNA gyrase, CS7.4, and the cold shock response in *Escherichia coli*. J. Bacteriol. 174: 5798-5802.

Jones, P.G., Mitta, M., Kim, Y., Jiang, W., and Inouye, M. 1996. Cold shock induces a major ribosomal-associated protein that unwinds double-stranded RNA in *Escherichia coli*. Proc. Natl. Acad. Sci. USA. 93: 76-80.

Jones, P.G., VanBogelen, R.A., and Neidhardt, F.C. 1987. Induction of proteins in response to low temperature in *Escherichia coli*. J. Bacteriol. 169: 2092-2095.

Kandror, O., and Goldberg, A.L. 1997. Trigger factor is induced upon cold shock and enhances viability of *Escherichia coli* at low temperatures. Proc. Natl. Acad. Sci. USA. 94: 4978-4981.

Kandror, O., Sherman, M., Moerschell, R., and Goldberg, A.L. 1997. Trigger factor associates with GroEL *in vivo* and promotes its binding to certain polypeptides. J. Biol. Chem. 272: 1730-1734.

Kandror, O., Sherman, M., Rhode, M., and Goldberg, A.L. 1995. Trigger factor is involved in GroEL-dependent protein degradation in *Escherichia coli* and promotes binding to GroEL to unfolded proteins. EMBO J. 14: 6021-6027.

Kashanchi, F., Duvall, J.F., Dittmer, J., Mireskandari, A., Reid, R.L., Gitlin, S.D., and Grady, J.N. 1994. Involvement of transcription factor YB-1 in human T-cell lymphotropic virus Type I basal gene expression. J. Virol. 68: 561-565.

Kumar, A., Malloch, R.A., Fujita, N., Smillie, D.A., Ishihama, A., and Hayward, R.S. 1993. The minus 35 region of *Escherichia coli* sigma 70 is inessential for initiation of transcription at an 'extended minus 10' promoter. J. Mol. Biol. 232: 406-418.

Landomery, M., and Sommerville, J. 1995. A role for Y-box proteins in cell proliferation. BioEssays. 17: 9-11.

LaTeana, A., Brandi, A., Falconi, M., Spurio, R., Pon, C.L., and Gualerzi, C.O. 1991. Identification of a cold shock transcriptional enhancer of the *Escherichia coli* gene encoding nucleoid protein H-NS. Proc. Natl. Acad. Sci. USA. 88: 10907-10911.

Lee, S.J., Xie, A., Jiang, W., Etchegaray, J.-P., Jones, P.G., and Inouye, M. 1994. Family of the major cold-shock protein, CspA (CS7.4) of *Escherichia coli*, whose members show a high sequence similarity with the eukaryotic Y-box binding proteins. Mol. Microbiol. 11: 833-839.

Lelivelt, M.J., and Kawula, T.H. 1995. Hsc66, an Hsp70 homolog in *Escherichia coli*, is induced by cold shock but not by heat shock. J. Bacteriol. 177: 4900-4907.

Lottering, E.A., and Streips, U.N. 1995. Induction of cold shock proteins in *Bacillus subtilis*. Curr. Microbiol. 30: 193-199.

Mackow, E.R., and Chang, F.N. 1983. Correlation between RNA synthesis and ppGpp content in *Escherichia coli* during temperature shifts. Mol. Gen. Genet. 192: 5-9.

Makhatadze, G.I., and Marahiel, M.A. 1994. Effect of pH and phosphate ions on self-association properties of the major cold-shock protein from *Bacillus subtilis*. Protein Sci. 3: 2144-2147.

Marr, A.G., and Ingraham, J.L. 1962. Effects of temperature on the composition of fatty acids in *Escherichia coli*. J. Bacteriol. 84: 1260-1267.

Matsumoto, K., and Wolffe, A.P. 1998. Gene regulation by Y-box proteins: coupling control of transcription and translation. Trends Cell Biol. 8: 318-323.

McGovern, V.P. and Oliver, J.D. 1995. Induction of cold-responsive proteins in *Vibrio vulnificus*. J. Bacteriol. 177: 4131-4133.

Michel, V., Labadie, J., and Hébraud, M. 1996. Effect of different temperature upshifts on protein synthesis by the psychrotrophic bacterium *Pseudomonas fragi*. Curr. Microbiol. 33: 16-25.

Michel, V., Lehoux, I., Depret, G., Anglade, P., Labadie, J., and Hebraud, M. 1997. The cold shock response of the psychrotrophic bacterium *Pseudomonas fragi* involves four low-molecular-mass nucleic acid-binding proteins. J. Bacteriol. 179: 7331-7342.

Mitta, M., Fang, L., and Inouye, M. 1997. Deletion analysis of *cspA* of *Escherichia coli*: requirement of the AT-rich UP element for *cspA* transcription and the downstream box in the coding region for its cold shock induction. Mol. Microbiol. 26: 321-335.

Mizushima, T., Kataoka, K., Ogata, Y., Inoue, R., and Sekimizu, K. 1997. Increase in negative supercoiling of plasmid DNA in *Escherichia coli* exposed to cold shock. Mol. Microbiol. 23: 381-386.

Morita, M.T., Tanaka, Y., Kodama, T., Kyogoku, Y., Yanagi, H., and Yura, T. 1999. Translational induction of heat shock transcription factor σ^{32}: evidence for a built-in RNA thermosensor. Genes Dev. 13: 655-665.

Moss, E.G., Lee, R.C., and Ambros, V. 1997. The cold shock domain protein LIN-28 controls developmental timing in *C. elegans* and is regulated by the *lin-4* RNA. Cell. 88: 637-646.

Nakashima, K., Kanamaru, K., Mizuno, T., and Horikoshi, K. 1996. A novel member of the *cspA* family of genes that is induced by cold-shock in *Escherichia coli*. J. Bacteriol. 178: 2994-2997.

Newkirk, K., Feng, W., Jiang, W., Tejero, R., Emerson, S.D., Inouye, M., and Montelione, G.T. 1994. Solution NMR structure of the major cold shock protein (CspA) from *Escherichia coli*: identification of a binding epitope for DNA. Proc. Natl. Acad. Sci. USA. 91: 5114-5118.

Nishiyama, H., Higashitsuji, H., Yokoi, H., Ito, K., Danno, S., Matsuda, T., and Fujita, J. 1997. Cloning and characterization of human CIRP (cold inducible RNA-binding protein) cDNA and chromosomal assignment of the gene. Gene. 204: 115-120.

Ohmori, M., Shimura, H., Shimura, Y., and Kohn, L.D. 1996. A Y-box protein is a suppressor factor that decreases thyrotropin receptor gene expression. Mol. Endocrinol. 10: 76-89.

Panoff, J.-M., Corroler, D., Thammavongs, B., and Boutibonnes, P. 1997. Differentiation between cold shock proteins and cold acclimation proteins in a mesophilic Gram-positive bacterium, *Enterococcus faecalis* JH2-2. J. Bacteriol. 179: 4451-4454.

Panoff, J.-M., Legrand, S., Thammavongs, B., and Boutibonnes, P. 1994. The cold shock response in *Lactococcus* subsp. *lactis*. Curr. Microbiol. 29: 213-216.

Phan-Thanh, L., and Gormon, T. 1995. Analysis of heat and cold shock proteins in *Listeria* by two-dimensional electrophoresis. Electrophoresis. 16: 444-450.

Ranjan, M., Tafuri, S., and Wolffe, A.P. 1993. Masking mRNA from translation in somatic cells. Genes Dev. 7: 1725-1736.

Roberts, M.E., and Inniss, W.E. 1992. The synthesis of cold shock proteins and cold acclimation proteins in the psychrophilic bacterium *Aquaspirillum arcticum*. Curr. Microbiol. 25: 275-278.

Ross, W., Gosink, K.K., Salomon, J., Igarashi, K., Zou, C., Ishihama, A., Severinov, K., and Gourse, R.L. 1993. A third recognition element in bacterial promoters: DNA binding by the α subunit of RNA polymerase. Science. 262: 1407-1413.

Rudd, K.E. 1998. Linkage map of *Escherichia coli* K-12, edition 10: the physical map. Microbiol. Mol. Biol. Rev. 62: 985-1019.

Sakamoto, T., and Bryant, D.A. 1997. Temperature-regulated mRNA accumulation and stabilization for fatty acid desaturase genes in the cyanobacteriim *Synechococcus* sp. strain PCC 7002. Mol. Microbiol. 23: 1281-1292.

Sato, N., and Nakamura, A. 1998. Involvement of the 5'-untranslated region in cold-regulated expression of the *rbpA1* gene in the cyanobacterium *Anabaena variabilis* M3. Nucleic Acids Res. 26: 2192-2199.

Schindelin, H., Jiang, W., Inouye, M., and Heinemann, U. 1994. Crystal structure of CspA, the major cold shock protein of *Escherichia coli*. Proc. Natl. Acad. Sci. USA. 91: 5119-5123.

Schindelin, H., Marahiel, M.A., and Heinemann, U. 1993. Universal nucleic acid-binding domain revealed by crystal structure of the *B. subtilis* major cold-shock protein. Nature. 364: 164-168.

Schnuchel, A., Wiltscheck, R., Czisch, M., Herrler, M., Willimsky, G., Graumann, P., Marahiel, M.A., and Holak, T.A. 1993. Structure in solution of the major cold-shock protein from *Bacillus subtilis*. Nature. 364: 169-171.

Scholz, C., Stoller, G., Zarnt, T., Fischer, G., and Schmid, F.X. 1997. Cooperation of enzymatic and chaperone functions of trigger factor in catalysis of protein folding. EMBO J. 16: 54-58.

Schröder, K., Graumann, P., Schnuchel, A., Holak, T.A., and Marahiel, M.A. 1995. Mutational analysis of the putative nucleic acid-binding surface of the cold-shock domain, CspB, revealed an essential role of aromatic and basic residues in binding of single-stranded DNA containing the Y-box motif. Mol. Microbiol. 16: 699-708.

Shinozaki, K., and Yamaguchi-Shinozaki, K. 1996. Molecular responses to drought and cold stress. Curr. Opin. Biotechnol. 7: 161-167.

Sinensky, M. 1974. Homeoviscous adaptation - a homeostatic process that regulates the viscosity of membrane lipids in *Escherichia coli*. Proc. Natl. Acad. Sci. USA. 71: 522-525.

Sprengart, M.L., Fatscher, H.P., and Fuchs, E. 1990. The initiation of translation in *E. coli*: apparent base pairing between the 16S RNA and downstream sequences of the mRNA. Nucleic Acids Res. 18: 1719-1723.

Sprengart, M.L., Fuchs, E., and Porter, A.G. 1996. The downstream box: an efficient and independent translation initiation signal in *Escherichia coli*. EMBO J. 15: 665-674.

Stoller, G., Rucknagel, K.P., Nierhaus, K.H., Schmid, F.X., Fischer, G., and Rahfeld, J.U. 1995. A ribosome-associated peptidyl-prolyl *cis/trans* isomerase identified as the trigger factor. EMBO J. 14: 4939-4948.

Stoller, G., Tradler, T., Rucknagel, K.P., Rahfeld, J.U., and Fischer, G. 1996. An 11.8 kDa proteolytic fragment of the *E. coli* trigger factor represents the domain carrying the peptidyl-prolyl *cis/trans* isomerase activity. FEBS Lett. 384: 117-122.

Storz, G. An RNA thermometer. 1999. Genes Dev. 13: 633-636.

Tanabe, H., Goldstein, J., Yang, M., and Inouye, M. 1992. Identification of the promoter region of the *Escherichia coli* major cold shock gene, *cspA*. J. Bacteriol. 174: 3867-3873.

Taura, T., Kusukawa, N., Yura, T., and Ito, K. 1989. Transient shut-off of Escherichia coli heat shock protein synthesis upon temperature shift down. Biochem. Biophys. Res. Commun. 163: 438-443.

Thieringer, H.A., Jones, P.G., and Inouye, M. 1998. Cold shock and adaptation. BioEssays. 20: 49-57.

Ting, J.P.Y., Painter, A., Zeleznik-Le, N.J., MacDonald, G., Moore, T.M., Brown, A., and Schwartz, B.D. 1994. YB-1 DNA binding protein represses interferon gamma activation of class II major histocompatibility complex genes. J. Exp. Med. 179: 1605-1611.

VanBogelen, R.A., and Neidhardt, F.C. 1990. Ribosomes as sensors of heat and cold shock in *Escherichia coli*. Proc. Natl. Acad. Sci. USA. 87: 5589-5593.

Wang, N., Yamanaka, K., and Inouye, M. 1999. CspI, the ninth member of the CspA family of *Escherichia coli*, is induced upon cold shock. J. Bacteriol. 181: 1603-1609.

Whyte, L.G., and Inniss, W.E. 1992. Cold shock proteins and cold acclimation proteins in a psychrotrophic bacterium. Can. J. Microbiol. 38: 1281-1285.

Willimsky, G., Bang, H., Fischer, G., and Marahiel, M.A. 1992. Characterization of *cspB*, a *Bacillus subtilis* inducible cold shock gene affecting cell viability at low temperatures. J. Bacteriol. 174: 6326-6335.

Yamanaka, K., Fang, L., and Inouye, M. 1998. The CspA family in *Escherichia coli*: multiple gene duplication for stress adaptation. Mol. Microbiol. 27: 247-255.

Yamanaka, K., and Inouye, M. 1997. Growth-phase-dependent expression of *cspD*, encoding a member of the CspA family in *Escherichia coli*. J. Bacteriol. 179: 5126-5130.

Yamanaka, K., Mitani, T., Ogura, T., Niki, H., and Hiraga, S. 1994. Cloning, sequencing, and characterization of multicopy supressors of a *mukB* mutation in *Escherichia coli*. Mol. Microbiol. 13: 301-312.

Yura, T., Nagai, H., and Mori, H. 1993. Regulation of the heat-shock response in bacteria. Annu. Rev. Microbiol. 47: 321-350.

Zou, Y., and Chien, K.R. 1995. EF1A/YB-1 is a component of cardiac HF-1A binding activity and positively regulates transcription of the myosin light-chain 2V gene. Mol. Cell. Biol. 15: 2972-2982.

Dedicated to Rudolf K. Thauer in celebration of his 60th birthday

3

Cold Shock Response in *Bacillus subtilis*

Peter L. Graumann, and Mohamed A. Marahiel

Biochemie, Fachbereich Chemie,
Hans-Meerwein-Straße, Philipps-Universität Marburg,
35032 Marburg, Germany

Abstract

Following a rapid decrease in temperature, the physiology of *Bacillus subtilis* cells changes profoundly. Cold shock adaptation has been monitored at the level of membrane composition, adjustment in DNA topology, and change in cytosolic protein synthesis/composition. Some major players in these processes (cold-stress induced proteins and cold acclimatization proteins, CIPs and CAPs) have been identified and mechanisms in cold shock acclimatization begin to emerge; however, important questions regarding their cellular function still need to be answered.

Introduction

Bacillus subtilis has the ability to slip into another skin when times get rough. Upon deprivation of nutrients or slow dehydration, the gram positive bacterium differentiates into a highly resistant spore (Stragier and Losick, 1996). This process takes about 7 h under optimal conditions. Being a common mesophilic soil bacterium, however, *B. subtilis* must be able to respond more rapidly to environmental changes, such as sudden and pronounced changes in temperature or osmolarity (Kempf and Bremer, 1998). *B. subtilis* is mainly found in the surface layers of the soil, where conditions change frequently and transiently. Thus, the physiology of the cells must adjust rapidly for growth and survival under different conditions, and adaptation after a decrease in temperature is discussed in this review.

In general, mesophilic bacteria have to cope with several recognized problems that arise following cold shock: a) membrane fluidity is too low, b) superhelical density of the DNA is too high for opening of the double helix, c) enzyme activities decrease profoundly, but probably to different extents, so protein levels must be adjusted, d) protein folding may be too slow or inefficient, e) ribosomes must be adapted to function properly at low temperatures, and f) secondary structures in RNA affect initiation of translation (Jones and Inouye, 1994; Graumann and Marahiel, 1996; Panoff et al., 1998). Upon a sudden decrease in temperature, B. subtilis performs a series of cellular adaptations, which have been monitored in all cellular fractions.

Membrane Adaptation

At 37°C, branched and straight-chain membrane lipids are fully saturated in B. subtilis to ensure the integrity of the membrane (Kaneda, 1991; Grau and de Mendoza, 1993). At lower temperatures, membrane fluidity must increase in order to avoid transtition from a liquid crystalline into a gel-like phase state of the lipid bilayer. To achieve a decrease in phospholipid membrane melting temperature, the ratio of anteiso-to iso-branched fatty acids in B. subtilis is dramatically increased (Klein et al., 1999). In E. coli, however, unsaturated fatty acids (UFAs) are synthesized in greater quantity at lower temperatures by a constitutive cytosolic enzyme, ACP synthase II, which is more active at lower temperatures (Cronan and Rock, 1996). In contrast, in B. subtilis the synthesis of a membrane desaturase that oxidizes phospholipids in the membrane is induced following cold shock (Figure 1; Aguilar et al., 1998). About 1 h after cold shock, transcription of the desaturase gene (des) is transiently induced, reaching 10 to 15-fold higher levels after 4 h. However, deletion of the des gene does not cause a detectable phenotype after cold shock (Aguilar et al., 1998). Recent data throw light on this behaviour, by showing that the anteiso-branched fatty acids and not UFAs to be the major fraction after temperature down shift in B. subtilis. In a defined minimal medium it has been shown that cold shock adaptation of B. subtilis depends on the presence of isoleucine (Ile) or precursors of anteiso-branched chain fatty acids, and that the branching pattern of membrane fatty acids (FA) switches from iso-focused (B. subtilis membranes contain a high proportion of branched chain FA, for review see, Kaneda, 1991) to anteiso-dominated after rapid cooling from 37°C to 15°C (Klein et al., 1999). Thus, an Ile-dependent change in the FA-branching profile appears to be the main mechanism for cold shock adaptation of the membrane in B. subtilis.

Chromosome Adjustment

After a temperature decrease, negative supercoiling is increased in the DNA of B. subtilis (Grau et al., 1994) and a variety of procaryotes, including hyperthermophilic archaea (Lopez-Garcia and Forterre, 1997). It is thought that underwinding of supercoiled DNA facilitates unwinding of the DNA duplex by helicases during replication and by RNA polymerase at lower temperatures. Artificial inhibition of gyrase activity (which introduces neg. supercoil) prevents the cold shock-induced decrease in linking number in B. subtilis (Grau et al., 1994). In E. coli, both subunits of gyrase are cold stress-induced proteins (CIPs) (Jones and Inouye, 1994), so increased synthesis

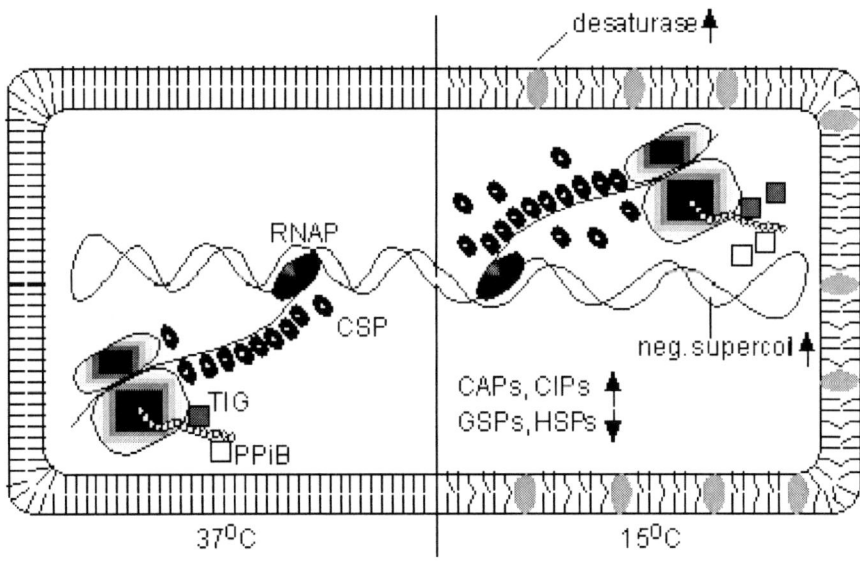

Figure 1. Temperature Adaptation. Model for adaptations following temperature shift down from 37°C to 15°C in *B. subtilis* and other mesophilic bacteria. For details, see text.

of this enzyme appears to account for the change in DNA topology. Inactivation of DNA gyrase also prevents cold dependent accumulation of UFAs (Grau *et al.*, 1994; see above), so transcriptional activation of *des* may depend on the topological state of the DNA. Possibly, the *des* promoter is only active when a certain threshold of negative supercoiling is reached.

Our unpublished results show that the nucleoid in *B. subtilis* is more condensed after cold shock than during exponential growth at 37°C (P. Graumann, unpublished), similar to nucleoids during stationary phase (Koch, 1996). This change in nucleoid structure may be due to a decrease in the cellular transcription/translation capacity. It has been proposed that decondensation of the nucleoid in growing cells is achieved by the coupling of synthesis of membrane proteins and secreted proteins to their incorporation into or transport across membranes. Thus the sites of active DNA transcription are pulled to the membranes (Woldringh *et al.*, 1995).

Cytosolic Response

In the cytosol, protein synthesis changes markedly after cold shock (Lottering and Streips, 1995; Graumann *et al.*, 1996; Figure 1). While synthesis of the majority of proteins decreases, a subset of CIPs is transiently induced, with a peak at about 1 h after temperature decrease from 37°C to 15°C. Thereafter, synthesis of most proteins resumes and induction of CIPs declines, such that 2 h after cold shock a new steady state of protein synthesis is reached (Graumann *et al.*, 1996). This pattern of synthesis is different from that at 37°C. Therefore, through a transient response, the cells adapt to the lower temperature with an adjustment of protein composition.

Table 1. Identified Cold Shock Stress Induced Proteins (CIPs) and CAPs in Bacillus subtilis

Protein	Name	Class	Function
Transcription/Translation			
CspB	cold shock protein B	CSI	initiation of translation, RNA-chaperone (?)
CspC	cold shock protein C	CSI	initiation of translation, RNA-chaperone (?)
CspD	cold shock protein D	CSI	initiation of translation, RNA-chaperone (?)
Translation			
S6	ribosomal protein S6	CSI	translation, folding of 16S RNA?
L7/L12	ribosomal protein L7/L12	CSI/CAP	translation, interaction with EFTu/aa-tRNA
L10	ribosomal protein L10	CAP	translation, interaction with EFTu/aa-tRNA
EFTs	GTP/GDP exchange factor	CAP	elongation of translation
EFTu	elongation factor	CAP	elongation of translation
EF-G	elongation factor	CAP	elongation of translation
Protein Folding			
PPiB	peptidylprolyl cis/trans isomerase	CSI	protein chaperone
Tig	trigger factor	CAP	protein chaperone (associated with ribosome)
General Metabolism			
CysK	cysteine synthase	CSI	amino acid synthesis
IlvC	ketolacid reductoisomerase	CSI	amino acid synthesis (Val, Ile)
GlnA	glutamine synthase	CAP	amino acid synthesis (Gln)
LeuC	betaisopropyl malate DH	CAP	amino acid synthesis (Leu)
ThrC	threonine synthase	CAP	amino acid synthesis (Thr)
AroF	chorismate synthase	CAP	amino acid synthesis
Gap	gyceraldehyde phosphate DH	CSI	glycolysis
TIM	triosephosphate isomerase	CSI	glycolysis
Fba	fructosebisphosphate aldolase	CSI	glycolysis
GuaB	inosine monophosphate DH	CAP	nucleotide synthesis
Chemotaxis			
CheY		TIP	regulation of flagellar rotation
Carbohydrate uptake			
Hpr		USP	PEP:phosphotransferase system
Iron Uptake			
DHBA	enterochelin synthase	CSI	synthesis of enterochelin (Fe-chelator)
Other			
Srf4	surfactin synthase 4, thioesterase	CSI	surfactin (fatty acid ?) synthesis
Csi12P		CSI	phosphorylated at initiation of sporulation
SpoVG		CSI	sporulation (septum formation)
Scp1		SCP	kinase-domain ?
Scp2		SCP	homology to Te-resistance from Alcaligenes
PspB	phage-shock protein B	CIP	membrane-associated (?) stress protein
Csi4b		CSI	unknown
Csi5		CSI	unknown
Csi9	RisB or AtpE	CSI	riboflavin synthase or E-ATPase
Csi15		CSI	unknown

CSI, cold stress induced protein (transient induction only after cold shock)
CAP, cold acclimatization protein (higher amount at cold temperatures)
SCP, salt and cold stress induced protein
TIP, temperature (cold and heat shock-) induced protein
USP, universal stress protein
DH, dehydrogenase

Although some proteins are induced in response to cold shock and salt stress or heat shock (SCP or TIP, see Table 1), generally, the synthesis of CSPs decreases following heat shock, and that of heat shock and general stress proteins (HSPs and GSPs) after cold shock (Graumann et al., 1996; see Figure 2). Likewise, heat shock and cold shock response are mutually exclusive in E. coli (Jones and Inouye, 1994).

Although the change in the pattern of protein synthesis is profound, the overall variation in protein composition is only moderate, because overall protein synthesis is much lower - only about 15-20% - after transition from 37°C to 15°C. 3 h after cold shock, the concentration of most CIPs is at least < 2 fold higher, and that of others, e.g. GSPs and many vegetative proteins, is lower (Figure 2). On the other hand, the amount of some proteins that are not strongly cold induced is also higher: probably through a slight but steady increase in synthesis, proteins like EFTs, EFTu and GlnA are present in a higher concentration 3 h after cold shock (Figure 2). These proteins, as opposed to CIPs, are termed cold acclimatisation proteins (CAPs, Table 1).

Through reverse genetics, the identity of several members of the cold shock stimulon has been determined (Graumann et al., 1996; Figure 2 and Table 1). The function of some CIPs is intriguing: like in E. coli, several CIPs/CAPs are associated with translation, and may adapt the ribosome to function at lower temperature. PPiB and TigBs in B. subtilis are folding catalysts that act as prolyl-isomerases (see below), and CspB-D may be RNA chaperones (see below). Several CIPs are involved in intermediary metabolism (Table 1), and may simply be needed at a higher level because they may be more inefficient enzymes at low temperatures than others in the corresponding biochemical pathway. CheY is phosphorylated by CheA, and regulates the direction of fagellar rotation; HPr is involved in the PTS system. Both proteins may be diffusion-limited factors in chemotaxis and sugar uptake, respectively, and therefore be induced to compensate for lower diffusion after cold shock. The function of CIPs such as Srf4 or PspB, however, is still a mystery.

Genes encoding CIPs are not clustered, but are distributed over the B. subtilis chromosome. Their induction after cold stress is probably achieved at the post-transcriptional level. A cold shock-like response can be induced upon treatment of B. subtilis with a low concentration of a translational inhibitor (Graumann et al., 1997); the same effect has previously been reported for E. coli (Van Bogelen and Neidhardt, 1990). In addition, cold shock induction of cspB transcription is less than 2-fold, but the synthesis of CspB strongly increases even when the cspB gene is under control of a constitutive promoter (Graumann et al., 1997). Induction of CIPs may therefore be achieved by a predominant translation of an existing pools of mRNA, which was also proposed for E. coli (Jones and Inouye, 1996). After cold shock, impaired initiation of translation (Jones and Inouye, 1996) may allow only the limited synthesis of CIPS, and resumption of general protein synthesis due to the adaptation of the translational machinery.

Protein Folding at Low Temperature: A Major Role for Prolyl Isomerases?

Peptidyl-prolyl bonds in proteins can not rotate freely, and have to exist either in a *cis* or a *trans* configuration. Isomerisation of these two states is therefore an important rate limiting step in the folding of many protein (Göthel and Marahiel, 1999). This reaction is catalysed by an ubiquitous class of enzymes called prolyl cis/trans isomerases. Two such enzymes are known in *B. subtilis*: PPiB and trigger factor (TigBs). Synthesis of both enzymes increases after cold shock (Figure 1, Graumann *et al.*, 1996), as was also

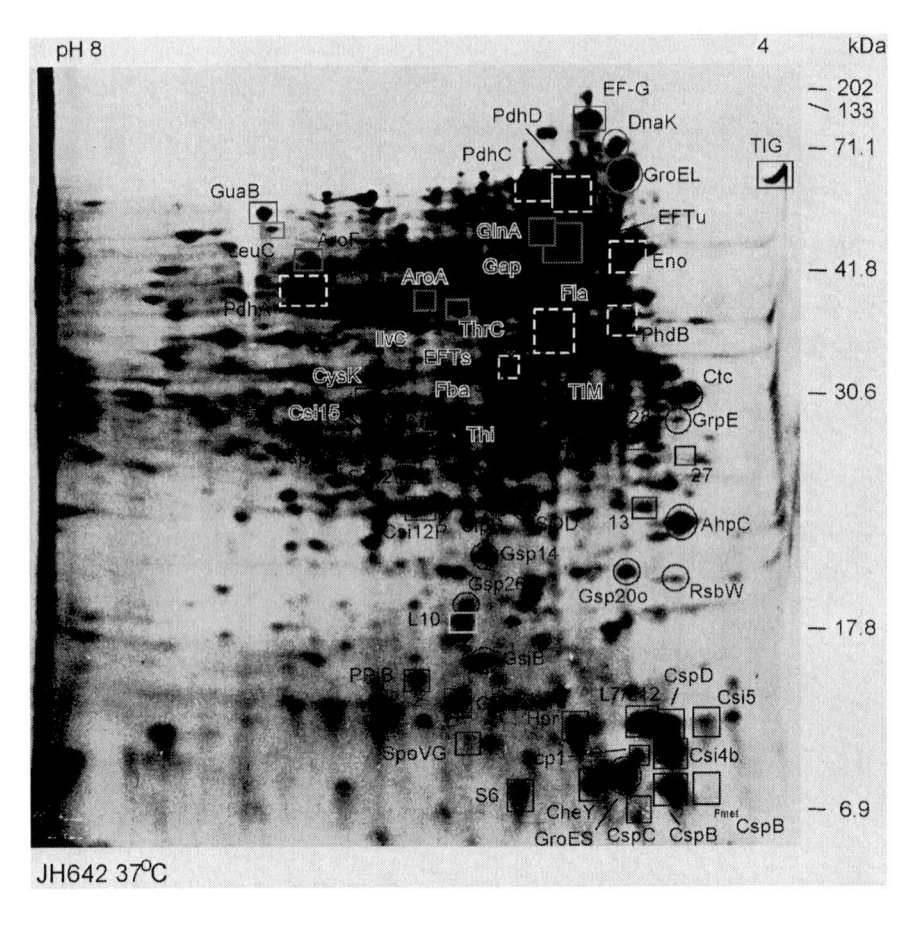

Figure 2. Stress Induction of Proteins. Coomassie stained second dimension SDS-PAGE 2D gels with 700 µg of soluble proteins from *B. subtilis* cells A) grown at 37°C during mid-exponential phase, and B) 3 h after cold shock to 15°C. HSPs are induced after heat shock, and function as protein chaperones; GSPs are induced after ethanol, salt and other stress (SOD: superoxide dismutase, AhpC: katalase, ClpP: protease, other GSPs have no known function; Bernhardt *et al.*, 1997); vegetative proteins are not stress inducible, except for TCA enzymes after glucose starvation (Fla: flagellin, PdhA-D: subunits of pyruvate dehydrogenase, Eno: enolase; Schmid *et al.*, 1997). Identification of protein spots (see Table 1 for CIPs/CAPs) from (Graumann *et al.*, 1996), (Antelmann *et al.*, 1997), Swiss-Prot accession numbers: Csi4B - P81094, Csi5 - P81095, Scp1 - P81099, Scp2 - P81100. Note that although some proteins are cold stress induced according to pulse-labelling experiments, their levels do not change visibly in Coomassie stained gels.

found for *E. coli* Tig protein (Kandror and Goldberg, 1997). Interestingly, an *E. coli tig* mutant was found to be cold sensitive (Kandror and Goldberg, 1997), whereas a *B. subtilis tig/ppiB* double mutant was impaired in stress adaptation in minimal medium (Göthel *et al.,* 1998). Thus, prolyl isomerization appears to be a committing step in protein folding at low temperatures, in contrast to heat shock conditions, where general misfolding and aggregation appear to be defective. In agreement with this, the synthesis of GroEL and GroES protein chaperones (HSPs) is reduced after cold shock (Graumann *et al.,* 1996).

Cold Shock Proteins: Important Function Not Only After Cold Stress

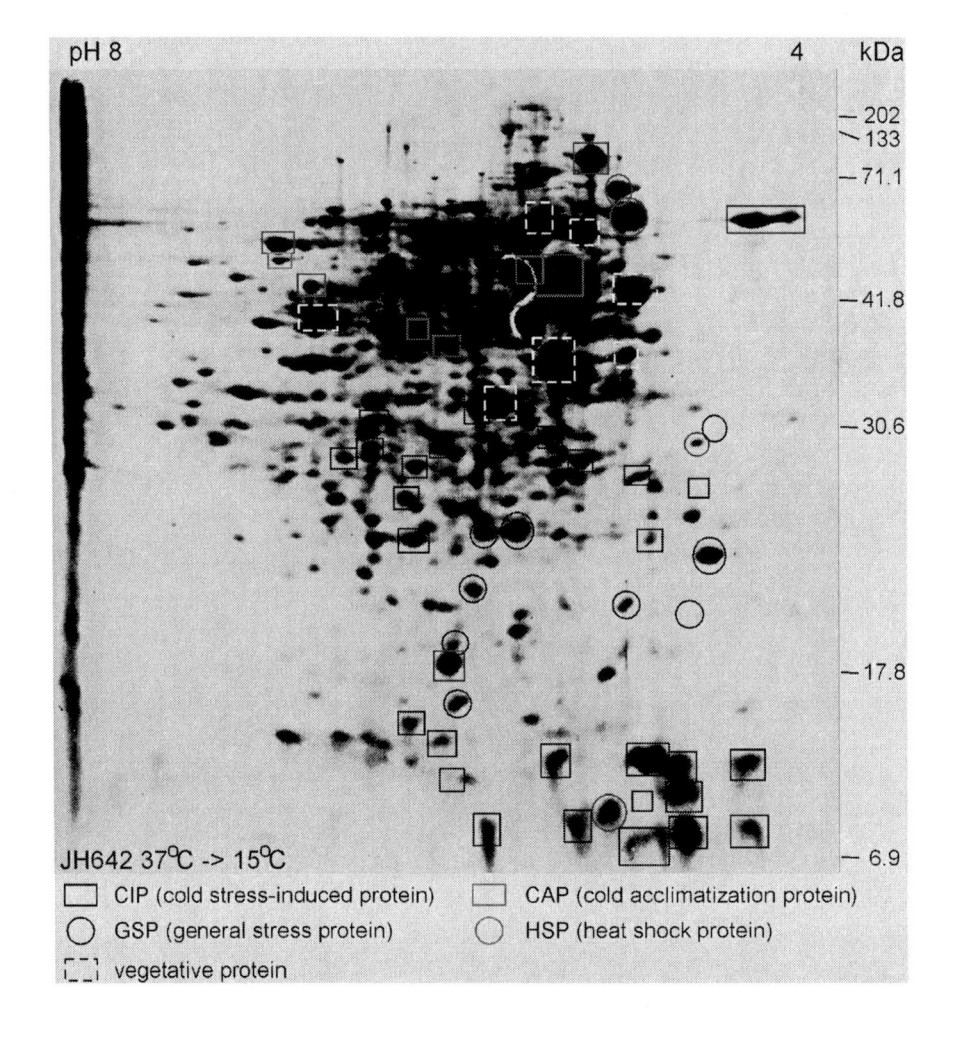

Cold shock proteins (CSPs) are the most strongly induced CIPs in *E. coli* and *B. subtilis*, as well as in a variety of other eubacteria. CSPs are small (7-7.5 kDa) proteins, highly conserved between even distantly related bacterial branches, and exist in families of up to 9 members (Graumann and Marahiel, 1998; Yamanaka et al., 1998). There are 3 csp genes (B, C and D) in B. subtilis, and all three gene products are cold inducible, while only three out of nine csp genes in E. coli are cold inducible. On the other hand, E. coli cspD is induced in response to starvation and stationary phase, but not after cold shock (Yamanaka et al., 1998). In Lactococcus sp., four CSPs are cold stress induced, while a fifth gene is constitutively expressed (Wouters et al., 1998). Thus, members of the CSP families can be differentially regulated. Interestingly, CspB and CspC, but not CspD, are also major stationary phase induced proteins in B. subtilis (Graumann and Marahiel, 1999), revealing that synthesis of CSPs can be increased in response to different stresses. Interestingly, rapid inactivation of ribosomes occurs after cold shock as well as after entry into stationary phase (Jones and Inouye, 1996; Wada, 1998), and may be a common trigger for CSP induction.

Deletion of any B. subtilis csp gene does not result in a detectable phenotype at physiological temperatures. A cspB null mutant has been shown to be sensitive to direct freezing from 37°C and a subsequent thawing. This phenotype could be partly restored by a pre-cold shock adaptation at 15°C prior to freezing (Willimsky et al., 1992). However, deletion of cspB and cspC or cspD results in a defect in cold stress acclimatization, and growth at 15 as well as 37°C. A cspC/cspD mutant is only defective in growth at low temperatures. These results show that CspB performs the most important function, and is complemented by CspC (mainly at low temperatures) and CspD (since a cspB/D mutant shows a stronger defect at 37°C than a cspB/C mutant, and vice versa at 15°C). Deletion of all three csp copies is only possible when cspB is present and induced in trans; in the absence of CSP production, B. subtilis is unable to grow, even at 37°C, which reveals an essential function of CSPs under optimal growth conditions. Since CspB and CspD are stable proteins in vivo (CspC becomes stable under cold shock conditions, Schindler et al., 1999), these results show that an increase in CSP synthesis and concentration is necessary for cold stress adaptation. Moreover, a cspB/C double mutant shows cell lysis after entry into stationary phase, indicating that an increase in synthesis of CspB and CspC (see above, Graumann and Marahiel, 1999) is also important for adaptation.

Function of CSPs as RNA Chaperones

It is clear that CSPs function at the level of transcription and/or translation, their exact molecular mode of function still needs to be elucidated. CSPs share a common fold, a five stranded β-barrel, first descibed for CspB of B. subtilis (Schindelin et al., 1993; Schnuchel et al., 1993; Figure 3). They carry conserved RNP1- and 2 motifs that are essential for binding to ssDNA and RNA (Schröder et al., 1995) on a β-sheet surface composed of three antiparrallel β-strands (Figure 3). Aromatic side chains in the RNP motifs are exposed to the solvent, an unusual feature in proteins, as hydrophobic residues are usually buried in the core of a protein. Since CspB can be photo-crosslinked to even short ssDNA molecules (Schröder et al., 1995), the rings of the aromatic residues are thought to stack with bases of ligand

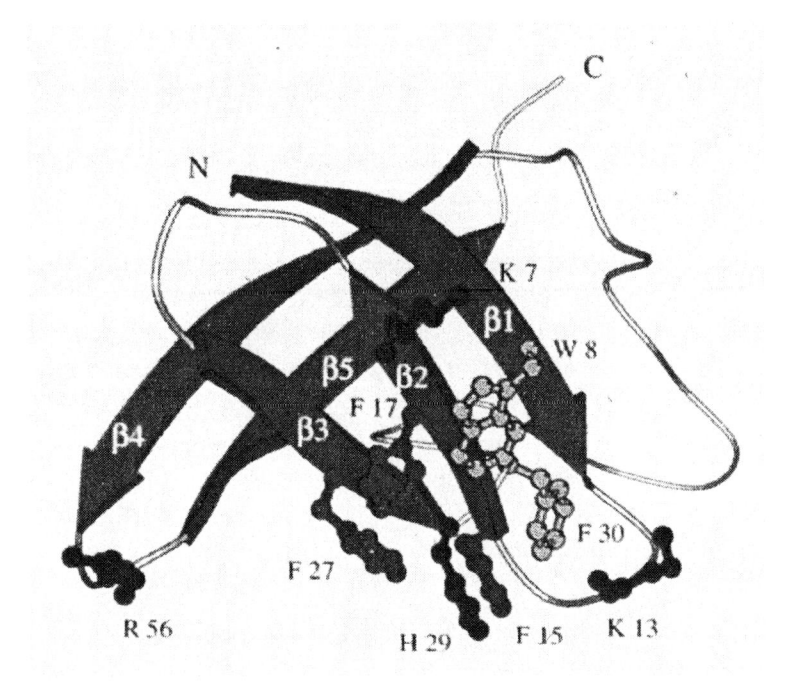

Figure 3. Structure of CspB from *B. subtilis* according to X-ray crystallography (Schindelin *et al.,* 1993). RNP1 motif (KGFGFIEV) and RNP2 motif (VFVH) are situated on β2 and β3 strands, respectively, and are essential for ssDNA and RNA-binding. Aromatic side chains are exposed to the solvent and are thought to stack with DNA/RNA bases, while basic residues are important for binding affinity by charge attraction.

nucleic acids. This is possible, since glycine residues in RNP1 allow a close approach of bases to the protein surface. Likewise, valine side chains in RNP2 are oriented towards the interior of the protein, allowing for space between aromatic RNP side chains (Figure 3). Additionally, basic charges are clustered around the RNP motifs (Figure 3), which create an attraction potential for nucleic acids, while the remaining CspB surface is highly negatively charged. The importance of ionic interaction is underlined by the finding that binding of some nucleic acids is strongly affected by the salt concentration (Lopez *et al.,* 1999).

Intriguingly, the aromatic side chains of RNP residues Phe15, Phe17, and Phe27 are not only important for nucleic acid binding, but also for protein stability, because their mutation strongly reduces conformational stability of CspB (Schindler *et al.,* 1998). Thus, there is no compromise between function and stability in the active site of Csps, but evolution has found an optimal structure for both features.

Although CSPs have a preference for sequences such as ATTGG (Graumann and Marahiel, 1994) and polyT (Lopez *et al.,* 1999), they bind rather non-specifically and cooperatively to single stranded nucleic acids (Graumann *et al.,* 1997). Affinity to RNA increases with increasing length of the substrate, but binding to longer (> 25 nucleotides) molecules requires that the RNA is devoid of secondary structures. CspA from *E. coli* has been shown to possess RNA-chaperone activity *in vitro* (Jiang *et al.,* 1997), which has led to the model that CSPs bind to nascent mRNA during transcription and prevent the formation of secondary structures that would inhibit initiation of translation. After cold shock, a higher concentration of CSPs may be needed to counterbalance increased stability of intramolecular basepairing in mRNA. On the other hand, affinity of CSPs to RNA is rather low (µmolar-range, Graumann *et al.,* 1997), consequently may be allowing the ribosomes to displace CSPs and initiate translation on a linear template. This model has recently received substantial support. Following photocrosslinking of nascent RNA to proteins in active transcription complexes, CspE was found to be a major RNA-bound constituent in *E. coli* (Hanna and Liu, 1998). CspE was heavily crosslinked to a short RNA of about ten nucleotidees in length only when this RNA was associated to the transcription complex. Moreover, *B. subtilis* CSPs were found to be stable proteins *in vivo*, in contrast to their high susceptibility to proteolytic degradation *in vitro* (CspB folds extremely rapidly with a low kinetic barrier towards unfolding, therefore about 1% of all molecules are present in an unfolded state in solution, Schindler *et al.,* 1995). However, addition of a substoichiometric amount of nucleic acid ligand strongly protected the CSPs against protease attack (Schindler *et al.,* 1999), which suggests that in the cell, CSPs are predominantly complexed most likely with mRNA.

Intriguingly, coupling of transcription to translation was also shown to be performed by the eukaryotic Y-box proteins, which contain a domain (cold shock domain) that is highly conserved to CSPs (Graumann and Marahiel, 1998; Matsumoto and Wolffe, 1998). Recently, a structural fold similar to CSPs was found in domain(s) within the S1 (Bycroft *et al.,* 1997), and IF1 ribosomal proteins of *E. coli* (Sette *et al.,* 1997) as well as in eukaryotic/archaeal eIF-5A factor (Kim *et al.,* 1998), all of which are involved in initiation of translation. Their common structure, called OB (oligomer-binding)-fold, has therefore been adapted during evolution to perform a variety of tasks in RNA/ribosome interactions.

The *csp* genes in *B. subtilis* appear to be autoregulated. Deletion of one or two genes results in higher synthesis of the remaining CSP(s). On the other hand, CspB can not be overproduced from additional copies of *cspB* on a plasmid (Graumann *et al.,* 1997). Tight control of CSP levels may be important, because moderate induction of *B. subtilis* CspB in *E. coli* at 37°C leads to a strong decrease in growth rate and a change in the pattern of protein synthesis (not seen after induction of a CspB mutant impaired in RNA-binding, Graumann and Marahiel, 1997). The highest affinity of *B. subtilis* CSPs was found for a sequence at the 5' end of their untranslated leader regions (5' UTRs). Possibly, CSPs bind tightly to their 5' UTRs and thereby reduce translation of *csp* mRNA.

Cold Shock Response in Other *Bacilli*

Cold shock from 65°C to 45°C induces a transient decrease in colony forming units in cultures of the thermophilic *Bacillus stearothermophilus* for 3 h, followed by resumption of exponential growth (Wu and Welker, 1991). During the adaptation period, the synthesis of several membrane and cytosolic proteins was induced. The presence of a translational inhibitor was found to increase the length of the adaptation suggesting that a change in protein synthesis is needed for cold stress adaptation. Interestingly, heat shock suppressed the synthesis of CIPs, while cold shock suppressed HSP production. Thus, thermophilic *bacilli* appear to have an analogous cold shock response to that of *B. subtilis*.

On the other end, the psychrophilic bacterium *B. cereus* was shown to contain at least 6 members of the Csp family, one of them was strongly induced after cold shock (Mayr *et al.*, 1996). CSPs seem to be an ubiquitous class of proteins, they are present in psychrophilic, mesophilic and thermophilic *bacilli* (Schröder *et al.*, 1993). As expected, Csps from thermophilic species are thermodynamically more stable than their counterparts from mesophilic *B. subtilis* (Perl *et al.*, 1998; Müller *et al.*, 2000). The molecular basis of this gain in stability has recently been investigated through determination of the three dimensional structure of CspB from *B. caldolyticus* (Bc-Csp), which is virtually identical to *B. subtilis* CspB in the central β-sheet and in the binding region for nucleic acids (Müller *et al.*, 2000). However, Bs-Csp possesses additional electrostatic interactions - located on its surface - that increase the free energy of stabilization.

It will be interesting to find out if they are generally major CIPs in *bacilli*, and if they perform a similar, essential function at low and optimal temperatures. Increased demand for RNA-chaperone activity at lower temperatures may be a recurring theme in many organisms: recently, a small RNA-binding protein (CIRP) was identified as the first cold shock protein in mammalian cells - including humans (Nishiyama *et al.*, 1997).

Acknowledgements

We would like to than Eva Uhlemann and Roland Schmid for their cooperation in microsequencing of protein spots. Work in the MAM laboratory was supported by the Deutsche Forschungsgemeinschaft, the Human Frontier in Science Program, and the Fonds der Chemischen Industrie.

References

Aguilar, P.S., Cronan, J.E., and de Mendoza, D. 1998. A *Bacillus subtilis* gene induced by cold shock encodes a membrane phospholipid desaturase. J. Bacteriol. 180: 2194-2200.

Antelmann, H., Bernhardt, J., Schmid, R., Mach, H., Völker, U., and Hecker, M. 1997. First steps from a two-dimensional protein index towards a response-regulation map for *Bacillus subtilis*. Electrophoresis. 18: 1451-1463.

Bernhardt, J., Völker, U., Völker, A., Antelmann, H., Schmid, R., Mach, H., and Hecker, M. 1997. Specific and general stress proteins in *Bacillus subtilis* - A two dimensional protein electrophoretic study. Microbiol. 143: 999-1017.

Bycroft, M., Hubbard, T.J.P., Proctor, M., Freund, S.M.V., and Murzin, A.G. 1997. The solution structure of the S1 RNA binding domain: a member of an ancient nucleic acid-binding fold. Cell. 88: 235-242.

Cronan, J. E., and Rock, C.O. 1996. Biosynthesis of membrane lipids. In: *Escherichia coli* and *Salmonella typhimurium*: Cellular and Molecular Biology, 2nd ed. F. C. Neidhardt, R. curtiss III, J. L. Ingram, E. C. C. Lin, K. B. Low, B. Magasanik, W. S. Reznikoff, M. Riley, M. Schaechter, and H. E. Umbarger, eda. ASM Press, Washington D.C. p. 612-636.

Göthel, S.F. and Marahiel, M.A. 1999. Peptidyl-prolyl cis-trans isomerases, a superfamily of ubiquitous folding catalysts. CMLS, Cell. Mol. Life Sci. 55: 423-436.

Göthel, S.F., Scholz, C., Schmid, F.X., and Marahiel, M.A. 1998. Cyclophilin and Trigger factor from *Bacillus subtilis* catalyze *in vitro* protein folding and are necessary for viability under starvation conditions. Biochem. 37: 13392-13399.

Grau, R., and de Mendoza, D. 1993. Regulation of the synthesis of unsaturated fatty acid in *Bacillus subtilis*. Mol. Microbiol. 8: 535-542.

Grau, R., Gardiol, D., Glikin, G.C., and de Mendoza, D. 1994. DNA supercoiling and thermal regulation of unsaturated fatty acid synthesis in *Bacillus subtilis*. Mol. Microbiol. 11: 933-941.

Graumann, P., and Marahiel, M.A. 1994. The major cold-shock protein of *Bacillus subtilis* CspB binds with high affinity to the ATTGG- and CCAAT sequences in single stranded oligonucleotides. FEBS Lett. 338: 157-160.

Graumann, P., and Marahiel, M.A. 1997. Effects of heterologous expression of CspB from *Bacillus subtilis* on gene expression in *Escherichia coli*. Mol. Gen. Genet. 253: 745-752.

Graumann, P., Schröder, K., Schmid, R., and Marahiel, M.A. 1996. Identification of cold shock stress induced proteins in *Bacillus subtilis*. J. Bacteriol. 178: 4611-4619.

Graumann, P., Wendrich, T.M., Weber, M.H.W., Schröder, K., and Marahiel, M.A. 1997. A family of cold shock proteins in *Bacillus subtilis* is essential for cellular growth and for efficient protein synthesis at optimal and low temperatures. Mol. Microbiol. 25: 741-756.

Graumann, P.L., and Marahiel, M.A. 1999. Cold shock proteins CspB and CspC are major stationary phase induced proteins in *B. subtilis*. Arch. Microbiol. 171: 135-138.

Graumann, P.L., and Marahiel, M.A. 1998. A superfamily of proteins containing the cold shock domain. Trends Biochem. Sci. 23: 286-290.

Hanna, M.M., and Liu, K. 1998. Nascent RNA in transcription complexes interacts with CspE, a small protein in *E. coli* implicated in chromatin condensation. J. Mol. Biol. 282: 227-239.

Jiang, W., Hou, Y., and Inouye, M. 1997. CspA, the major cold-shock protein of *Escherichia coli*, is an RNA chaperone. J. Biol. Chem. 272: 196-202.

Jones, P.G., and Inouye, M. 1994. The cold shock response - a hot topic. Mol. Microbiol. 11: 811-818.

Jones, P.G., and Inouye, M. 1996. RbfA, a 30S-ribosomal binding factor, is a cold shock protein whose absence triggers the cold shock response.

Mol. Microbiol. 21: 1207-1218.

Kandror, O., and Goldberg, A.L. 1997. Trigger factor is induced upon cold shock and enhances viability of *Escherichia coli* at low temperatures. Proc. Natl. Acad. Sci. USA 94: 4978-4981.

Kaneda, 1991. Iso- and antiso-fatty acids in bacteria: Biosynthesis, function and taxonomic significance. Microbiol. Rev. 55: 288-302

Kempf, B., and Bremer, E. 1998. Uptake and synthesis of compatible solutes as microbial stress responses to high-osmolarity environments. Arch. Microbiol. 170: 319-330.

Klein, W., Weber, M.H.W., and Marahiel, M.A. 1999. The cold shock response of *Bacillus subtilis*: An isoleucine dependent switch in the fatty acids branching pattern for membrane adaptation to low temperature. J. Bacteriol. 181: 5341-5349.

Kim, K.K., Hung, L.W., Yokota, H., Kim, R., and Kim, S.H. 1998. Crystal structure of eukaryotic translation initiation factor 5A from *Methanococcus jannaschii* at 1.8 A resolution. Proc. Natl. Acad. Sci. USA. 95: 10419-10424.

Koch, A.L. 1996. What size should a bacterium be? A question of scale. Annu. Rev. Microbiol. 50: 317-348.

Lopez, M.M., Yutani, K., and Makhatadze, G.I. 1999. Interactions of the major cold shock protein of *Bacillus subtilis* CspB with single stranded DNA templates of different base composition. J. Biol. Chem. 274: 33601-33608.

Lopez-Garcia, P., and Forterre, P. 1997. DNA topology in hyperthermophilic archaea: reference states and their variation with growth phase, growth temperature, and temperature stresses. Mol. Microbiol. 23: 1267-1279.

Lottering, E.A., and Streips, U.N. 1995. Induction of cold shock proteins in *Bacillus subtilis*. Current Microbiol. 30: 193-199.

Matsumoto, K., and Wolffe, A.P. 1998. Gene regulation by Y-box proteins: coupling control of transcription and translation. Trends Cell Biol. 8: 318-323.

Mayr, B., Kaplan, T., Lechner, S., and Scherer, S. 1996. Identification and purification of a family of dimeric major cold shock protein homologs from the psychrotrophic *Bacillus cereus* WSBC 10201. J. Bacteriol. 178: 2916-2925.

Müller, U., Perl, D., Schmid, F.X., Heinemann, U. 2000. Thermal stability and atomic-resolution crystal structure of the *Bacillus caldolyticus* cold shock protein. J. Mol. Biol. 297: 975-988.

Nishiyama, H., Higashitsuji, H., Yokoi, H., Itoh, K., Danno, S., Matsuda, T., and Fujita, J. 1997. Cloning and characterization of human CIRP (cold inducible RNa-binding protein) cDNA and chromosomal assignment of the gene. Gene. 204: 115-120.

Panoff, J.-M., Thammavongs, B., Gueguen, M., and Boutibonnes, P. 1998. Cold stress responses in mesophilic bacteria. Cryobiol. 36: 75-83.

Perl, D., Welker, C., Schindler, T., Schröder, K., Marahiel, M.A., Jaenicke, R., and Schmid, F.X. 1998. Conservation of rapid two-state folding in mesophilic, thermophilic and hyperthermophilic cold shock proteins. Nature Struct. Biol. 5: 229-235.

Schindelin, H., Marahiel, M.A., and Heinemann, U. 1993. Universal nucleic acid-binding domain revealed by crystal structure of the *B. subtilis* major cold-shock protein. Nature. 364: 164-167.

Schindler, T., Graumann, P.L., Perl, D., Ma, S., Schmid, F.X., and Marahiel, M.A. 1998. The family of cold shock proteins of *Bacillus subtilis*. Stability and dynamics in vitro and in vivo. J. Biol. Chem. 274: 3407-3413.

Schindler, T., Herrler, M., Marahiel, M.A., and Schmid, F.X. 1995. Extremely rapid protein folding in the absence of intermediates. Nature Struc. Biol. 2: 663-673.

Schmid, R., Bernhardt, J., Antelmann, H., Völker, A., Mach, H., Völker, U., and Hecker, M. 1997. Identification of vegetative proteins for a two-dimensional protein index of *Bacillus subtilis*. Microbiol. 143: 991-998.

Schnuchel, A., Wiltscheck, R., Czisch, M., Herrler, M., Willimsky, G., Graumann, P., Marahiel, M.A., and Holak, T.A. 1993. Structure in solution of the major cold-shock protein from *Bacillus subtilis*. Nature. 364: 169-171.

Schröder, K., Graumann, P., Schnuchel, A., Holak, T.A., and Marahiel, M.A. 1995. Mutational analysis of the putative nucleic acid binding surface of the cold shock domain, CspB, revealed an essential role of aromatic and basic residues in binding single-stranded DNA containing the Y-box motif. Mol. Microbiol. 16: 699-708.

Schröder, K., Zuber, P., Willimsky, G., Wagner, B., and Marahiel, M.A. 1993. Mapping of the *Bacillus subtilis cspB* gene and cloning of its homologs in thermophilic, mesophilic and psychotropic bacilli. Gene. 136: 277-280.

Sette, M., van Tilborg, P., Spurio, R., Kaptein, R., Paci, M., Gualerzi, C.O., and Boelens, R. 1997. The structure of the translational initiation factor IF1 from *E. coli* contains an oligomer-binding motif. EMBO J. 16: 1436-1443.

Stragier, P., and Losick, R. 1996. Molecular genetics of sporulation in *Bacillus subtilis*. Annu. Rev. Genet. 30: 297-341.

Van Bogelen, R., and Neidhardt, F.C. 1990. Ribosomes as sensors of heat and cold shock in *Escherichia coli*. Proc. Natl. Acad. Sci. USA. 87: 5589-5593.

Wada, A. 1998. Growth phase coupled modulation of *Escherichia coli* ribosomes. Genes Cells. 3: 203-208.

Willimsky, G., Bang, H., Fischer, G., and Marahiel, M.A. 1992. Characterization of *cspB*, a *Bacillus subtilis* inducible cold shock gene affecting viability at low temperatures. J. Bacteriol. 174: 6326-6335.

Woldringh, C.L., Jensen, P.R., and Westerhoff, H.V. 1995. Structure and partitioning of bacterial DNA: determined by a balance of compaction and expansion forces? FEMS Microbiol. Lett. 131: 235-242.

Wouters, J.A., Sanders, J.W., Kok, J., de Vos, W.M., Kuipers, O.P., and Abee, T. 1998. Clustered organization and transcriptional analysis of a family of five *csp* genes of *Lactococcus lactis* MG1363. Microbiol. 144: 2885-2893.

Wu, L., and Welker, N.E. 1991. Temperature-induced protein synthesis in *Bacillus stearothermophilus* NUB36. J. Bacteriol. 173: 4889-4892.

Yamanaka, K., Fang, L., and Inouye, M. 1998. The CspA family in *Escherichia coli*: multiple gene duplication for stress adaptation. Mol. Microbiol. 27: 247-256.

4

Cold Acclimation and Cold Shock Response in Psychrotrophic Bacteria

Michel Hébraud[1], and Patrick Potier[2]

[1]Unité de Recherches sur la Viande, Equipe Microbiologie,
Institut National de la Recherche Agronomique de Theix,
63122 Saint-Genès Champanelle, France
[2]Laboratoire d'Ecologie Microbienne,
Unité Mixte de Recherche du CNRS 5557,
Université Claude Bernard, Lyon I,
69622 Villeurbanne Cedex, France

Abstract

Psychrotrophic bacteria are capable of developing over a wide temperature range and they can grow at temperatures close to or below freezing. This ability requires specific adaptative strategies in order to maintain membrane fluidity, the continuance of their metabolic activities, and protein synthesis at low temperature. A cold-shock response has been described in several psychrotrophic bacteria, which is somewhat different from that in mesophilic micro-organisms: (i) the synthesis of housekeeping proteins is not repressed following temperature downshift and they are similarly expressed at optimal and low temperatures (ii) cold-shock proteins or Csps are synthesized, the number of which increases with the severity of the shock (iii) a second group of cold-induced proteins, *i.e.* the cold acclimation proteins or Caps, comparable with Csps are continuously synthesized during prolonged growth at low temperature. Homologues to CspA, the major cold-shock protein in *E. coli*, have been described in various psychrotrophs, but unlike their mesophilic counterparts, they belong to the group of Caps. Although they have been poorly studied, Caps are of particular importance since they differentiate psychrotrophs from mesophiles, and they are probably one of the key determinant that allow life at very low temperature.

Introduction

In nature, many bacteria can grow harmoniously in very hostile environments such as polar regions and cold water, acidic hot springs, salterns, dry rock surfaces, deserts, or at depth in the sea. Such organisms are able to withstand harsh conditions, and they are often submitted to rapid variations of the environment. Among the various environmental factors that condition the viability of microorganisms, their growth and physiology, temperature is of particular interest since it affects immediately the interior of the cells. According to their ability to grow at high, intermediate or low temperature, microorganisms have been divided into three broad categories: thermophiles, mesophiles and psychrophiles, respectively. The last category has been further subdivided into psychrophiles *sensu stricto*, which have optimal growth temperatures below 15°C and an upper limit of 20°C, and psychrotrophs (psychrotolerants) which are able to divide at 0°C or below and grow optimally at temperatures around 20-25°C (Morita, 1975). Psychrophilic and psychrotrophic microorganisms are of particular importance in global ecology since the majority of terrestrial and aquatic ecosystems of our planet is permanently or seasonally submitted to cold temperatures: the world's oceans occupy 71% of the earth surface and 90% of their volume is below 5°C; the polar regions represent 14% of the earth surface and if one includes alpine soils and lakes, snow and icefields, fresh waters and caves, more than 80% of the earth biosphere is below 5°C. Microorganisms capable of coping with low temperatures are widespread in these natural environments where they often represent the dominant flora and they should therefore be regarded as the most successful colonizers of our planet (Russell, 1990). Furthermore, the development of the industrialised production of foods and the increased use of refrigeration for their long conservation, have greatly enhanced the importance of psychrotrophic bacteria. Their presence in foodstuffs is a frequent cause of spoilage and food poisoning.

Recently, the effects of hypothermic stress on the protein content of various microorganisms has been investigated: bacteria as well as eucaryotic cells respond to an abrupt decrease in temperature by overexpressing a specific set of proteins, the cold-shock proteins (Csps). It has been suggested that a universal cold shock regulon exists, and that the Csps expressed after an hypothermic stress are probably required for optimal adaptation to the lower temperature. Paradoxically, there is still a paucity of information concerning the cold-shock response in cold-adapted bacteria and its role in subsequent adaptation of cells to low temperatures, whereas it has been extensively studied in *Escherichia coli*, the paradigm of mesophilic bacteria. After a brief review of the physiology of cold-adapted microorganisms, we shall concentrate mainly on the cold-shock response in psychrotrophic bacteria which is different from that of mesophilic bacteria in several respects.

Physiology of Cold-Adapted Microorganisms

Psychrophiles and psychrotrophs belong to extremely diverse genera and, in addition to mechanisms developed during the course of evolution, adaptation to identical thermal constraints may imply common molecular mechanisms that allow the maintenance of vital cellular functions at low temperatures. In this respect, psychrotrophs are interesting since, while being able to grow at temperatures close to or below freezing, they kept their ability to withstand mild temperatures. Psychrotrophic microorganisms are more ubiquitous than psychrophilic bacteria and they are numerous (both quantitatively and in terms of number of species) even in permanently cold environments. In their natural habitats, they are frequently submitted to large and rapid temperature changes and they can develop over a wide temperature range (up to 40°C, whereas the temperature range that permits growth of most other bacteria does not exceed 30°C). This ability to cope with such temperature shifts must be accompanied by adaptative changes in response to alterations of numerous physical and biochemical parameters, including solubility, reaction kinetics, membrane fluidity, protein conformation and stability and changes in gene expression. Therefore, the biochemical effects allowing bacterial cells to adapt to wide temperature changes are likely to be complex, involving a number of interacting phenomena.

Kinetics of Growth at Low Temperature
Microbial growth is the result of a sequence of interrelated chemical reactions on which the effects of temperature can be expressed by the Arrhenius equation by plotting the log of the specific growth rate constant against the reciprocal of absolute temperature in degrees Kelvin. The linear descending portion of the curve corresponds to the physiologically normal temperature range for growth, and indicates the obedience of the growth rate to temperature on the Arrhenius principle regarding a chemical reaction. At temperatures above or below this range, a deviation from linearity occurs, and rather than just a continual decrease in growth rate, the curve finally becomes vertical and growth ceases. The general form of the Arrhenius curve is the same for all microorganisms, but cardinal temperatures (*i.e.* maximum, optimum and minimum) are lower for cold-adapted bacteria. For all microorganisms, temperatures outside the linear range of the Arrhenius plots are stress-inducing temperatures, and what remains to be explained is why, whereas deviation from linearity occurs at about 20°C for mesophilic bacteria, it occurs between 5 and 10°C for psychrotrophs and below 0°C for psychrophiles. Therefore, genotypic adaptation to very low temperatures may represent the sum of many end-points of phenotypic adaptations of cell structure/metabolism, which can be achieved to various extents by different microorganisms (Russell, 1990; Gounot, 1991; Gounot and Russell, 1999).

Lipid Composition and Membrane Fluidity at Low Temperature
One of the best studied effects of low temperatures on bacterial physiology concerns the mechanisms by which microorganisms maintain an optimal

degree of fluidity of their membrane, and there is abundant information concerning the lipid composition in psychrotrophs and psychrophiles (for reviews: Herbert R., 1986; Russell, 1990; Russell and Fukunaga, 1990). Depending upon the strain, membrane fluidity at low temperature can be achieved in several ways, such as increasing the ratio of unsaturated fattyacyl residues and/or *cis* double bonds, chain shortening, and in some rare cases, methyl branching (Shaw and Ingraham, 1967; McElhaney, 1982; Russell, 1989). This diversity of response reflects the fact that different fatty acid compositions can provide similar thermal properties, and no specific mechanism appears to be linked to psychrophily. Furthermore, the temperature-dependent changes in membrane composition of cold-adapted microorganisms are essentially the same as those observed in mesophiles. Russell (1990) suggested that the relevant factor that differentiates the adjustment of membrane fluidity in mesophiles and cold-adapted bacteria is the timescale of the adaptive changes after a sudden temperature decrease, which would be particularly important for growth in thermally unstable cold habitats where large and rapid thermal fluctuations exist. These temperature-triggered lipid changes can be mediated by (i) enzyme activation responsible for the modification of pre-existing lipids, and (ii) by *de novo* synthesis of specific enzymes following a temperature downshift. In this respect, the product of the *des* gene which is induced following cold-shock in *Bacillus subtilis* was recently identified as a membrane phospholipid desaturase (Aguilar *et al.*, 1998) and could be the only component of the desaturation system induced by the temperature (Aguilar *et al.*, 1999). It is therefore likely that similar cold-shock induced enzymes exist in cold-adapted microorganisms, which would be responsible for the rapid homeoviscous adaptation of their membrane.

Cold-Adapted Enzymes
In contrast to the situation in mesophiles, all structural and metabolic proteins of cold-adapted bacteria have to be functional at low temperature, sometimes near or below 0°C. Catalysis at low temperature is a thermodynamic challenge which raises a number of questions including, whether growth at low temperature involves the synthesis of new proteins, or whether the cellular proteins are sufficiently «heat and cold stable» to function normally at all temperatures at which growth is possible. Both situations seem to coexist within cold-adapted bacteria. Synthesizing more enzymes or synthesizing enzymes characterized by temperature-independent reaction rates (perfectly evolved enzymes) are two other strategies to maintain sustainable growth at low temperature (Feller and Gerday, 1997). The first possibility appears energetically expensive and cannot be extended to a whole organism, whereas perfectly evolved enzymes are quite rare.

Compared with their mesophilic counterparts, enzymes from cold-adapted bacteria are more thermolabile but they are much more active at low temperatures (Rentier-Delrue *et al.*, 1993; Gerike *et al.*, 1997; Choo *et al.*, 1998; Kulakova *et al.*, 1999). It is generally thought that cold-adapted enzymes have evolved toward a high conformational flexibility, which would be responsible for their increased catalytic efficiency at low temperature and their low thermal stability (Feller et al., 1997). The correlation between conformational flexibility and enzyme activity has been studied by comparing a few cold-adapted enzymes with their mesophilic homologues such as the β-lactamase from the Antarctic psychrophile *Psychrobacter immobilis* A5 (Feller et al., 1997), the α-amylase from *Alteromonas haloplanctis* (Feller et al., 1992; Aghajari et al., 1998), and the subtilisin from the Antarctic *Bacillus* TA41 (Davail et al., 1994) or their thermophilic homologues such as the malate dehydrogenase from *Aquaspirillium arcticum* (Kim et al., 1999). It appears that protein flexibility can be achieved by several means, including the reduction of electrostatic noncovalent weak interactions (salt bridges, polar interactions between aromatic side chains, hydrogen bonds) and the decrease of hydrophobicity. However, no general rule governs this adaptative strategy and each enzyme can gain flexibility by one or a combination of the above modifications (Feller and Gerday, 1997). The need for more flexible molecules in order to gain functionality at low temperature is probably not limited to proteins. For example, high levels of dihydrouridine have been found in the tRNA of three psychrotrophilic bacteria (ANT-300 and *Vibrio* sp. 5710 and 29-6), which are responsible for the maintenance of conformational flexibility and dynamic motion in RNA at low temperature (Dalluge et al., 1997). This, together with protein modifications, may provide an advantage in organisms growing under conditions where thermal motion, enzymatic reaction rates and intermolecular interactions of biomolecules are compromised.

Protein Content at Low Temperature

Another aspect of adaptation to cold concerns the synthesis at low temperature of a specific set of proteins that are not (or poorly) present at milder temperatures. This particular class of proteins, referred to as cold acclimation proteins (Caps), is permanently synthesized during continuous growth at low temperature. Such Caps have been described in a variety of phylogenetically unrelated cold-adapted bacteria (Potier et al., 1990; Araki, 1992; Roberts and Inniss, 1992; Whyte and Inniss, 1992; Hébraud et al., 1994; Berger et al., 1996) and they are likely to play a fundamental role for life in the cold. Therefore, the presence of Caps appears to be a general feature of cold-adapted microorganisms that differentiates them from mesophiles. These proteins may be involved in important metabolic function(s) at low temperature by maintaining membrane fluidity and/or by replacing cold-denatured peptides. A low-temperature-specific proteolytic system has been described for the psychrotrophic *Arthrobacter globiformis* SI 55 (Potier et al., 1987a, b), and some Caps could act as cold-specific proteases that eliminate denatured proteins whose accumulation would be deleterious for the cells.

Protein Synthesis at Low Temperature

The important points that remain to be explained is how protein synthesis proceeds in cold-adapted bacteria at temperatures not suitable for mesophiles and what are the factors that preclude protein synthesis in mesophiles at temperatures below 8-10°C. Szer (1970) isolated a protein (factor P) by washing ribosomes of the psychrotrophic *Pseudomonas* sp. 412. The washed ribosomes, while retaining activity at 25-37°C, largely lost their capacity to function at low temperature but this could be restored by addition of the protein washings. It has been established that protein synthesis is inhibited at 0°C in *E. coli* while all macromolecules necessary for translation are present and energy is still available (Das and Goldstein, 1968; Friedman *et al.*, 1969; Okuyama and Yamada, 1979). The ability to synthesize more protein is almost instantly restored after only a brief exposure to 37°C. The primary effect of cold results in a polysomal run-off and accumulation of 70S ribosomal particles (Das and Goldstein, 1968; Broeze *et al.*, 1978). These particles are then dissociated and subunits are accumulated at the expense of polysomes (Friedman *et al.*, 1969). Das and Goldstein (1968) proposed that the cold-labile step in protein synthesis was the reattachment of ribosome to mRNA after they had run-off the 3' end at the completion of the cycle of transcription. Broeze *et al.* (1978) found that the initiation of protein synthesis was much more resistant to a sudden decrease in temperature in the psychrotroph *Pseudomonas fluorescens* compared with *E. coli* showing that this early step in translation might be the one that is adapted in cold-loving bacteria.

Cold-Shock Response in Psychrotrophic Bacteria

Cold-shock is characterized by a sudden transfer of cells to lower temperatures. Although such rapid temperature downshifts are unlikely to occur in natural environments, they provide interesting laboratory situations that largely contribute to the elucidation of the molecular mechanisms by which cells respond to cold. Most of our knowledge on how cells respond to rapid temperature downshifts originates from studies on mesophilic microorganisms such as *E. coli* and *B. subtilis* (for review, see Jones and Inouye, 1994; Wolffe, 1995; Graumann and Marahiel, 1996; 1999; Thieringer *et al.*, 1998; Phadtare *et al.*, 1999; Yamanaka *et al.*, 1999). For these bacteria, a sudden temperature decrease to 10-15°C creates a stress situation to which cells respond by specific adaptative mechanisms which, to a certain extent, allow their survival and subsequent growth at the lower temperature. A similar cold stress also exists, but at lower temperatures, when psychrotrophic bacteria are shifted outside the linear range of their Arrhenius plot. The question which arises is what are the cold-shock specific adaptative mechanisms that allow psychrotrophs to cope with lower temperatures than mesophiles. Although the cold-shock response in psychrotrophic bacteria is still poorly documented, recent data indicate that it presents similarities but also differences which may provide a partial answer to this question.

Effect of Cold-Shock on Growth
Psychrotrophic bacteria respond to temperature downshifts by a lag period before growth resumes at a rate characteristic of the new temperature (Phan-

Thanh and Gormon, 1995; Bayles et al., 1996; Berger et al., 1996; Chasseignaux and Hébraud, 1997; Michel et al., 1997). The duration of the lag period increases with the magnitude of the downshift and/or with the lowering of the final shock temperature (Table 1). The same phenomena are also observed for E. coli (Ng et al., 1962; Jones et al., 1987) whereas for other mesophilic bacteria (Panoff et al., 1994; McGovern and Oliver, 1995; Lottering and Streips, 1995; Kim and Dunn, 1997) and for the psychrophilic Vibrio sp. (Araki, 1991a), no lag phase is observed and growth continues at an intermediate rate followed by a growth rate characteristic of the final low temperature. Obviously, different mechanisms exist for bacteria to respond to temperature downshifts, and the time to re-adapt to the low temperature is not directly dependent on the range of growth of the bacterium, i.e., of its psychrophilic, psychrotrophic, or mesophilic character.

Effect of Cold-Shock on Protein Synthesis
In mesophilic bacteria, cold-shock results in the transient inhibition of the synthesis of the bulk of proteins, the so-called housekeeping proteins. In B. subtilis, this inhibition is only partial and at least 75 proteins are permanently synthesized upon cold-shock (Graumann and Marahiel, 1996). The growth of B. subtilis continues at a reduced doubling time without apparent growth lag. The response of E. coli is somewhat different since the transient inhibition of housekeeping proteins is total and results in a growth lag period (Jones et al., 1987; Thieringer et al., 1998). One of the most significant difference between mesophiles and cold-adapted bacteria is that the relative rate of synthesis of most cytosolic proteins is maintained after cold-shock. This has been demonstrated in Vibrio sp. ANT-300, (Araki, 1991a, b) B. psychrophilus (Whyte and Inniss, 1992), A. globiformis (Berger et al., 1996) and P. fragi (Michel et al., 1997). Obviously, it can be assumed that, as opposed to mesophiles, regulatory factor(s) exists in cold-adapted bacteria prior to cold-shock that allows the maintenance of a functional translational machinery at low temperature. However, in spite of this continuous protein synthesis, growth of psychrotrophic bacteria ceases transiently after a cold-shock and additional regulatory mechanisms may exist which allow growth resumption at temperatures close to freezing.

An intermediate adaptative mechanism seems to exist in those pathogenic bacteria that are not psychrotrophic per se, but are capable of residual growth at low temperature (i.e. around 4°C). For example, Phan-Thanh and Gormon (1995) reported that the synthesis of half the proteins synthesized at 25°C is turned off while many others are depressed more than twofold in the first hours following a cold-shock to 4°C in Listeria monocytogenes and Listeria innocua. However, a lag phase is systematically observed which increases with the severity of the cold shock (Table 1). It is therefore likely that several strategies exist for adaptation to cold that may account for the continuum of lower growth temperatures within the microbial world.

Table 1. Duration of Lag-Phase According to the Magnitude of Temperature Downshifts

Micro-organisms	Cold-shocks [a]	Lag phase [b]	References
Pseudomonas fragi	20 → 5	3	Michel et al., 1997
	30 → 5	5	
Listeria monocytogenes LO28	15 → 5	1*	Chasseignaux and Hébraud, 1997
	20 → 5	3*	
	25 → 5	5*	
	30 → 5	> 11*	
	37 → 5	> 11*	
Listeria monocytogenes 10403S	37 → 10	2	Bayles et al., 1996
	37 → 5	>2	
Arthrobacter globiformis	25 15	1 - 2	Berger et al., 1996
	25 → 10	4 - 6	
	25 → 8	6 - 7	
	25 → 6	7 - 8	
	25 → 4	10 - 12	
Escherichia coli	37 → 10	4	Jones et al., 1992
	24 → 10	2	

[a] initial and final temperatures (°C).
[b] lag phase in hours except * (days).

Cold-Induced Proteins in Psychrotrophic Bacteria

The outstanding common feature of the microbial response to cold-shock is that all bacteria studied to date, including thermophiles (Schröder et al., 1993), mesophiles (Jones et al., 1987; Schröder et al., 1993; Jones and Inouye, 1994; McGovern and Oliver, 1995; Graumann et al., 1996), psychrotrophs (Cloutier et al., 1992; Whyte and Inniss, 1992; Schröder et al., 1993; Berger et al., 1996; Gumley and Inniss, 1996; Michel et al., 1997), and psychrophiles (Araki, 1991a, b; Roberts and Inniss, 1992), overexpress a specific subset of proteins: the cold-shock proteins (Csps). In psychrotrophic bacteria, a basal set of Csps exists which is overexpressed even after mild shocks, and additional Csps appear with more severe cold-shocks (Table 2). In *A. globiformis* SI 55, a mathematical analysis of protein appearance revealed that the sequence of cold-induced proteins expression following cold-shocks of increasing magnitude is not random, and the Csps fall into different groups: one is common to all shocks, one is specific of mild shocks, and one is specific of large shocks that occur outside the range of linearity of the Arrhenius plot (Berger et al., 1996). Hence, the number of Csps together with the magnitude of their induction depend upon the range of the temperature shift: the larger this range, the more pronounced the response (Table 2). It appears therefore that the synthesis of Csps in psychrotrophic bacteria results from active regulatory mechanisms and it is likely that some if not all these proteins are essential for helping cells to recover from the temperature downshift.

A general feature of cold-adapted bacteria is that the relative levels of cold-induced proteins are much lower than those reported in mesophiles

Table 2. Cold-Induced Proteins, Including Caps (in parentheses), and Common Csps between Cold-Shocks of increasing Magnitude, in *P. fragi* (Michel et al., 1997) and *A. globiformis* (Berger et al., 1996)

Micro-organisms	Cold-shocks [a]	Number of Csps	Number of common Csps [b]	Total number of Caps [c]
Pseudomonas fragi	20 → 5	15		
	30 → 5	25 (11 Caps)	12 (20 → 5)	20
Arthrobacter globiformis	25 → 15	13		
	25 → 10	14	12 (25 → 15)	
	25 → 8	17	12 (25 → 15)	
			13 (25 → 10)	
	25 → 6	19	12 (25 → 15)	
			13 (25 → 10)	
			17 (25 → 8)	
	25 → 4	26 (9 Caps)	11 (25 → 15)	18
			12 (25 → 10)	
			16 (25 → 8)	
			18 (25 → 6)	

[a] initial and final temperatures (°C).
[b] in parentheses, the cold-shock considered.
[c] determined during prolonged growth at 4°C.

(Araki, 1991a, b; Cloutier et al., 1992; Roberts and Inniss, 1992; Whyte and Inniss, 1992; Schröder et al., 1993; Berger et al., 1996; Gumley and Inniss, 1996; Michel et al., 1997). This is probably due to the fact that, as opposed to mesophiles in which Csps are the prevailing proteins that are synthesized following cold-shock, the overall synthesis of cellular proteins is not inhibited in cold-adapted bacteria. This diversion of the translational machinery may therefore account for the lower rates of cold-induced protein synthesis in psychrotrophs.

Although there are very few data on the overall cold-shock response in cold-adapted microorganisms, more Csps seem to be overexpressed than in mesophilic bacteria, such as *E. coli* (Jones et al., 1987; Goldstein et al., 1990; La Teana et al., 1991; Jones et al., 1992; Jones and Inouye, 1994; Lelivelt and Kawula, 1995; Jones et al., 1996) and *Lactococcus lactis* (Panoff et al., 1994). Julseth and Inniss (1990) reported the induction of 26 Csps after a 24 to 4°C cold-shock in the psychrotrophic yeast *Trichosporon pullulans*. Cloutier et al. (1992) showed that Arctic *rhizobium* strains respond to very low temperature (-10°C) by synthesizing more proteins than temperate strains do at higher temperatures. When the psychrotrophic bacteria *P. fragi* and *A. globiformis* are subjected to cold-shocks from their optimal growth temperature to 4-5°C, 25 and 26 Csps can be detected, respectively (Berger et al., 1996; Michel et al., 1997). It is important to underline that among the total number of proteins overexpressed upon cold-shock in these two bacteria, some behave like Caps and are still synthesized during prolonged growth at the shift temperature (Table 2). Therefore, the number of Csps *per se* in *P. fragi* and *A. globiformis* is similar to that reported for mesophilic bacteria. However, this simple numeric deduction is certainly not sufficient to say that a totally common pattern exists for Csps synthesis between these two thermal bacterial groups, and some of Csps in mesophiles are Caps in psychrotrophs (see below). Synthesizing large numbers of Csps is probably not a pre-requisite for adaptation to very low temperatures since *B. subtilis*, which is unable to grow at temperatures near or below freezing, was shown to synthesize 37 cold-shock proteins after temperature downshift from 37 to 15°C (Graumann *et al.*, 1996).

Kinetics of Expression of Cold-Induced Proteins

Studies with *A. globiformis* (Berger *et al.*, 1996) or *P. fragi* (Michel *et al.*, 1997) indicated that some cold-induced proteins appear very rapidly following temperature downshift. This number increases during post-shift to reach maximal induction before or at mid-lag phase. A similar situation exists in *E. coli* in which Csps are very quickly induced after a downshift from 37 to 10°C, with an optimal level of induction towards 2 hours (Goldstein *et al.*, 1990).

In *P. fragi*, the kinetic pattern of protein expression following cold-shocks from 30 or 20 to 5°C allows to differentiate the Csps and the Caps according to their transient or permanent overexpression, respectively (Michel *et al.*, 1997). Each of these two classes can be further divided in two subclasses according to their immediate or delayed overexpression. Induction of these late cold-induced proteins usually takes place one hour after the shift or later during the lag period. A similar pattern of expression exists in *A. globiformis*, in which the appearance of cold-induced proteins is sequential: early Csps and Caps are synthesized immediately following cold-shock, followed by late Csps and Caps, and finally growth resumes (Berger *et al.*, 1996). It can be speculated therefore that early cold-induced proteins are necessary for the expression of the late cold-induced proteins, both of them being essential for growth resumption at the low temperature (Figure 1). The persistence of some of these proteins (early and late Caps) during prolonged growth in the cold may be important to maintain a balanced physiology at low temperature and may constitute one of the key determinant of psychrotolerance.

The CspA-Like Proteins of Psychrotrophic Bacteria

All the recent studies concerning the cold-shock response in psychrotrophic bacteria are generally descriptive and most Csps and Caps have not been identified. This situation originates from the fact that these microorganisms are poorly studied, their genetics is totally unknown and, due to their relative low levels of synthesis, insufficient amounts of Csps and Caps are generally present on two-dimensional gel electrophoresis to allow micro-sequencing. Therefore, most of what we know so far about cold-inducible proteins in these bacteria comes from analogies with *E. coli* and *B. subtilis*, for which the nature of the majority of Csps has been identified. In particular, most studies concerning cold-induced proteins in psychrotrophs have focused on proteins that resemble CspA, the major cold-shock protein in *E. coli*. CspA is believed to play a central role in the cold-shock response in this bacterium. It is specifically produced after cold-shock and its production reaches

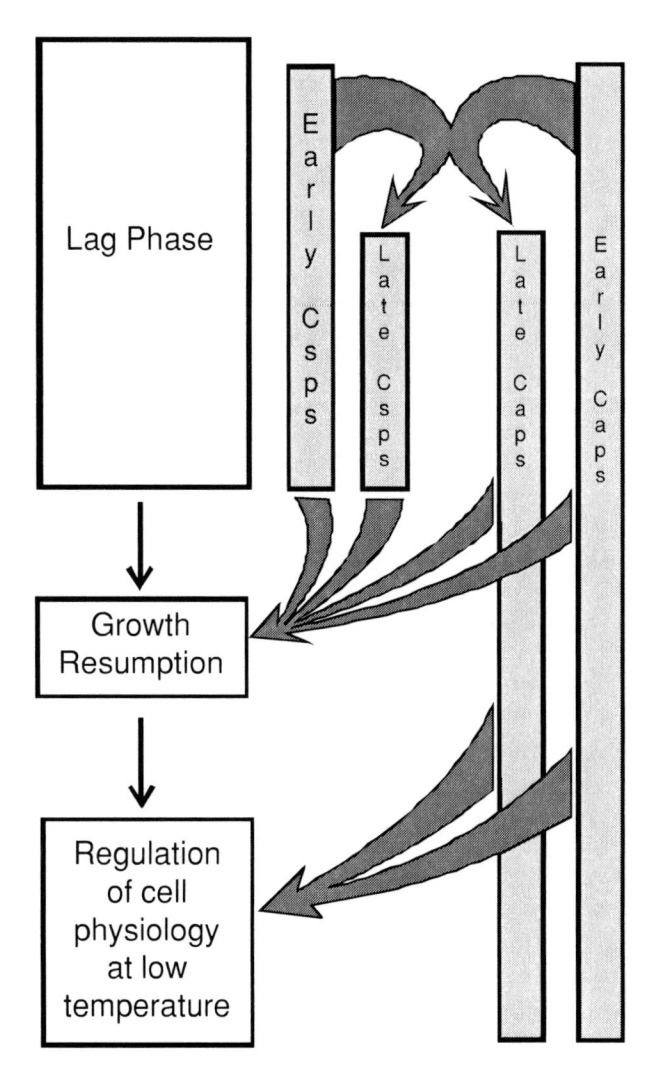

Figure 1. Schematic representation of the hypothetical interactions between cold-induced proteins and acclimation to low temperature in psychrotrophic bacteria.

13% of total cellular protein synthesis following temperature downshift. It belongs to a family of nine homologous proteins, CspA to CspI, of which only CspA, CspB, CspG and CspI are cold-shock inducible, but are differentially regulated (Etchegaray et al., 1996; Etchegaray and Inouye, 1999; Wang et al., 1999). Similar to E. coli, B. subtilis also contains a family of CspA-like proteins, and three homologous proteins (CspB, CspC and CspD) have been found to be essential for low temperature adaptation (Graumann et al., 1996). CspA homologues have been identified in a wide number of Gram negative and Gram positive eubacteria whatever their growth range of temperature (Av-Gay et al., 1992; Willimsky et al., 1992; Schröder et al., 1993; Hébraud et al., 1994; Ray et al., 1994; Berger et al., 1996; Graumann et al., 1996; Mayr et al., 1996; Chapot-Chartier et al., 1997; Francis and Stewart, 1997;

Mayo et al., 1997; Michel et al., 1997; Kim et al., 1998; Perl et al., 1998; Fujii et al., 1999; Neuhaus et al., 1999; Perrin et al., 1999; Stamm et al., 1999; Welker et al., 1999; Wouters et al., 1998; 1999; Yamanaka et al., 1999). However, most of these studies do not detail the low-temperature inducible or noninducible character of these genes.

Genes homologous to cspA have been found in several psychrotolerant bacteria including Listeria (Francis and Stewart, 1997) and Bacillus cereus (Mayr et al., 1996) and Yersinia enterocolitica (Neuhaus et al., 1999), as well as in psychrotrophic bacteria such as Arthrobacter (Ray et al., 1994; Berger et al., 1996), Pseudomonas (Hébraud et al., 1994; Ray et al., 1994; Michel et al., 1997), Bacillus globisporus (Schröder et al., 1993), and Micrococcus roseus (Ray et al., 1994). All these genes encode highly conserved proteins that share a high level of identity with all members of the CspA family in E. coli and with the homologous proteins in other mesophilic bacteria. No particular amino-acid sequence nor domain seems to exist in the CspA-like proteins of cold-adapted bacteria that could differentiate them from their mesophilic counterparts. They also contain the highly conserved RNP-1 and RNP-2 motifs, both domains being involved in the binding to RNA and single-stranded DNA, thereby suggesting related functions of these proteins.

The number of cspA-like genes varies in the cold-adapted bacteria. Only one gene is present in A. globiformis SI 55 (Berger et al., 1997; Berchet and Potier, unpublished) whereas at least 2, 4 and 6 genes exist in Y. enterocolitica (Neuhaus et al., 1999), P. fragi (Hébraud et al., 1994) and B. cereus WSBC 10201 (Mayr et al., 1996), respectively. The cspA-like genes found in A. globiformis SI 55, Y. enterocolitica and P. fragi are all cold-inducible. On the contrary, only one of the six cspA homologues in B. cereus WSBC 10201 is overexpressed at low temperature and following cold-shock from 30 to 7°C. In a collection of 100 strains of B. cereus belonging to different thermal groups, the presence of a cold-inducible cspA gene correlates with the ability of some strains to grow at or below 7°C (Francis et al., 1998). It is therefore likely that, as their mesophilic counterparts, the products of cold-inducible cspA-like genes play a major role in cell adaptation to low temperatures.

Similar to B. subtilis in which a cspB/cspC/cspD triple deletion mutation is lethal (Graumann et al., 1997), allelic exchange of the unique copy of the cspA-like gene with a deleted copy is lethal in A. globiformis SI 55 (Berchet and Potier, unpublished). This suggests a more general role of these genes than their sole implication in cold acclimation processes, and the presence of at least one copy is essential for survival.

Expression of CspA-Like Proteins in Psychrotrophic Bacteria

Expression of CspA-like proteins is an immediate response to temperature downshift in all the psychrotrophic bacteria studied to date. For example, they appear within the first hour postshift in P. fragi (Michel et al., 1997) and as soon as 20 minutes in A. globiformis SI 55 (Berger et al., 1997). In this latter bacterium, the transcripts of this gene pre-exist prior to cold-shock and post-transcriptional regulation is likely to account for its induction.

In *A. globiformis* SI 55 submitted to various cold shocks at temperatures inside the range of linearity of the Arrhenius plot, *i.e.* down to 6°C, synthesis of the CspA-like protein is transient and ceases between 2-4 h postshift (Berger *et al.*, 1996). However, when cells are transferred to a colder temperature, this protein is continuously synthesized at a similar level, not only during the lag phase, but also during prolonged growth at the shift temperature. Hence, this CspA-like protein in *A. globiformis* SI 55 was named CapA for «cold acclimation protein A». The particular continuous expression of CspA homologues at low temperature has also been described in *B. cereus* (Mayr *et al.*, 1996), and in *P. fragi* (Hébraud *et al.*, 1994). Therefore, the CspA-like protein in all these microorganisms can be referred to as an early Cap, and this may represent a common feature of cold-adapted bacteria that distinguishes them from mesophiles.

The four CspA homologues in *P. fragi* are the products of four independent genes, and their expression varies at different growth temperatures as well as following temperature shifts. All four proteins are overexpressed following a 30 or 20°C to 5°C shift: two of them (CapA and CapB) are Caps since they are still synthesized during prolonged growth at 5°C, whereas the two others cannot be considered as Caps *per se* since their level decreases slowly after growth resumption at 5°C (Michel *et al.*, 1997). Surprisingly, these two last CspA homologues are also heat-shock inducible (Michel *et al.*, 1996) and were therefore referenced as TapA and TapB for «temperature adaptation proteins A and B». In cultures at steady-state temperatures, CapA and CapB are optimally expressed between 4 to 10°C, whereas TapA and TapB are optimally expressed at 25 to 30°C, *i.e.* at temperatures around their optimal for growth (Hébraud *et al.*, 1994). The fact that either of these protein pairs is overexpressed according to the growth temperature means there is a subtle way of regulation with the probable involvement of temperature-sensitive molecules. Such a pattern of expression has not been described in other bacteria and further studies are necessary to gain insights on the precise role of these four small proteins in *P. fragi* at extremes of temperature.

Finally, an attempt to correlate CapA overexpression to cold adaptation was made in *A. globiformis* SI 55 (Berger *et al.*, 1997). A temporary block in protein synthesis during cold-shock treatment results in an additional lag that increases with the duration of the inhibition. CapA expression is delayed for similar periods of time and whatever the length of protein synthesis inhibition, growth resumption systematically occurs 12 to 14 h after CapA synthesis. Therefore, a positive correlation exists in *A. globiformis* SI 55 between CapA appearance and growth resumption at low temperature, but whether or not CapA synthesis is causal remains to be determined.

Coming in from the Cold?

For several decades, it was commonly thought that life on our planet appeared at high temperature, and the primitive character of hyperthermophiles has become a paradigm. The recent discovery of psychrophilic archaea

(DeLong et al., 1994; DeLong, 1998), which branch close to the base of the universal phylogenetic tree, revived the idea of a cold origin (Forterre, 1999). Therefore, the presence of Csps in thermophiles, mesophiles and psychrotrophs raises interesting questions about the warm or cold origin of their relevant genes. If the «warm hypothesis» is correct, it can be speculated that these genes, being involved in some vital processes, have been conserved throughout evolution from thermophiles to psychrotrophs. Caps would therefore be the manifestation of a recent evolutionary process by which some bacteria gained the ability to colonize cold habitats. According to the «cold hypothesis», a complete set of genes necessary for life at low temperature would have existed in MRCA, the so-called Most Recent Common Ancestor (Galtier et al., 1999), which would have been conserved in psychrotrophic and psychrophilic bacteria. Part of these genes, such as those encoding Caps, may have been lost during the course of evolution that led to the appearance of mesophilic microorganisms. If such is the case, perturbations of the translational machinery, the dramatic overexpression of a few Csps and the absence of Caps following cold-shock would be the manifestation of an incomplete and deregulated mechanism that prevents these cells from growing in the cold. Therefore, Csps and Caps could provide interesting models to gain an understanding on the origins of life on our planet.

Conclusions

Psychrotrophic bacteria can grow harmoniously at temperatures not suitable for mesophilic bacteria. In spite of their importance in natural environments and in refrigerated food, cold-adapted bacteria have been largely ignored and their biotechnological exploitation has remained untapped. This situation is changing and recent interest has focused on the biotechnological potential of psychrophilic and psychrotrophic microorganisms and their cold-active enzymes: expression of eucaryotic recombinant proteins of medical interest that are naturally thermosensitive (e.g. insulin, interferon, enkephalins, growth hormones, etc), production of cold-active enzymes (cellulases, amylases, lipases, pectinases, etc) that could be used in low-energy processes, soil or water bioremediation in cold environments. A better knowledge of the physiology and genetics of these microorganisms is necessary to control their degradative abilities in natural cold environments, to optimize the production of their enzymes for biotechnological purposes, or to prevent their growth in foods. The fundamental question of how they function at low temperature remains to be elucidated. There is probably no general and unique explanation for adaptation of psychrophilic and psychrotrophic bacteria to temperatures near or below freezing. Growth and cell multiplication at such temperatures underlie that these bacteria developed adaptative strategies in order to maintain enzyme activity and membrane functionality. Cold-shock inducible proteins are certainly involved in the mechanisms by which these cells overcome temperature downshift. Among them, cold acclimation proteins (Caps) are probably the specific determinants that differenciate psychrotrophs from mesophiles and allow continuous growth at very low temperature. Determination of the nature and of the role of these Caps is certainly the major task for further studies in order to gain an understanding in psychrophily.

References

Aghajari, N., Feller, G., Gerday, C., and Haser, R. 1998. Structures of the psychrophilic *Alteromomas haloplanctis* alpha-amylase give insights into cold adaptation at a molecular level. Structure. 6: 1503-1516.

Aguilar, P.S., Cronan, J.E. Jr, and de Mendoza, D. 1998. A *Bacillus subtilis* gene induced by cold shock encodes a membrane phospholipid desaturase. J. Bacteriol. 180: 2194-2200.

Aguilar, P.S., Lopez, P., and de Mendoza, D. 1999. Transcriptional control of the low-temperature-inducible *des* gene, encoding delta5 desaturase of *Bacillus subtilis*. J. Bacteriol. 181: 7028-7033.

Araki, T. 1991a. The effect of temperature shifts on protein synthesis by the psychrophilic bacterium *Vibrio* sp. Strain ANT-300. J. Gen. Microbiol. 137: 817-826.

Araki, T. 1991b. Changes in rates of synthesis of individual proteins in a psychrophilic bacterium after a shift in temperature. Can. J. Microbiol. 37: 840-847.

Araki, T. 1992. An analysis of the effect of changes in growth temperature on proteolysis *in vivo* in the psychrophilic bacterium *Vibrio* sp. strain ANT-300. J. Gen. Microbiol. 138: 2075-82.

Av-Gay, Y., Aharonowitz, Y., and Cohen, G. 1992. *Streptomyces* contain a 7.0 kDa cold shock like protein. Nucl. Acids Res. 20: 5478.

Bayles, D.O., Bassam, A.A., and Wilkinson, B.J. 1996. Cold stress proteins induced in *Listeria monocytogenes* in response to temperature downshock and growth at low temperatures. Appl. Environ. Microbiol. 62: 1116-1119.

Berger, F., Morellet, N., Menu, F., and Potier, P. 1996. Cold shock and cold acclimation proteins in the psychrotrophic bacterium *Arthrobacter globiformis* SI55. J. Bacteriol. 178: 2999-3007.

Berger, F., Normand, P., and Potier, P. 1997. *capA*, a *cspA*-like gene that encodes a cold acclimation protein in the psychrotrophic bacterium *Arthrobacter globiformis* SI55. J. Bacteriol. 179: 5670-5676.

Broeze, R., Solomon, C., and Pope, D. 1978. Effects of low temperature on *in vivo* and *in vitro* protein synthesis in *Escherichia coli* and *Pseudomonas fluorescens*. J. Bacteriol. 134: 861-874.

Chapot-Chartier, M.P., Schouler, C., Lepeuple, A.S., Gripon, J.C., and Chopin, M.C. 1997. Characterization of *cspB*, a cold-shock-inducible gene from *Lactococcus lactis*, and evidence for a family of genes homologous to the *Escherichia coli cspA* major cold shock gene. J. Bacteriol. 179: 5589-5593.

Chasseignaux, E., and Hébraud, M. 1997. Effect of cold shocks on growth and protein synthesis of *Listeria monocytogenes.* Purification of the main cold shock protein. In: Proceedings of the Euroconference on Microbial Response to Stress : What's New and Can it be Applied? I. Sá-Correia,

A.M. Fialho, and C.A. Viegas, eds. Gráfica Povoensa, Lda. Póvsa de Santa Iria, Portugal. p. 64.

Choo, D.W., Kurihara, T., Suzuki, T., Soda, K., and Esaki, N. 1998. A cold-adapted lipase of an Alaskan psychrotroph, *Pseudomonas* sp. strain B11-1: gene cloning and enzyme purification and characterization. Appl. Environ. Microbiol. 64: 486-491.

Cloutier, J., Prévost, D., Nadeau, P., and Antoun, H. 1992. Heat and cold shock protein synthesis in arctic and temperate strains of *Rhizobia*. Appl. Environ. Microbiol. 58: 2846-2853.

Dalluge, J.J., Hamamoto, T., Horikoshi, K., Morita, R.Y., Stetter, K.O., and McCloskey, J.A. 1997. Posttranscriptional modification of tRNA in psychrophilic bacteria. J. Bacteriol. 179: 1918-1923.

Das, H., and Goldstein, A. 1968. Limited capacity for protein synthesis at zero degrees centigrade in *Escherichia coli*. J. Mol. Biol. 31: 209-226.

Davail, S., Feller, G., Narinx, E., and Gerday, C. 1994. Cold adaptation of proteins. J. Biol. Chem. 269: 1-6.

DeLong, E., Wu, K.Y., Prézelin, B.B., and Jovine, R.V.M. 1994. High abundance of Archaea in Antarctic marine picoplankton. Nature. 371: 695-697.

DeLong, E.F. 1998. Everything in moderation: archaea as 'non-extremophiles'. Curr. Opin. Genet. Dev. 8: 649-654.

Etchegaray, J.P., and Inouye, M. 1999. CspA, CspB, and CspG, major cold shock proteins of *Escherichia coli,* are induced at low temperature under conditions that completely block protein synthesis. J. Bacteriol. 181: 1827-1830.

Etchegaray, J.P., Jones, P.G., and Inouye, M. 1996. Differential thermoregulation of two highly homologous cold-shock genes, cspA and cspB, of *Escherichia coli*. Genes Cells. 1: 171-178.

Feller, G., and Gerday, C. 1997. Psychrophilic enzymes: molecular basis of cold adaptation. Cell. Mol. Life Sci. 53: 830-841.

Feller, G., Lonhienne, T., Deroanne, C., Libioulle, C., VanBeeumen, J., and Gerday, C. 1992. Purification, characterization and nucleotide sequence of the thermolabile alpha-amylase from the Antarctic psychrotroph *Alteromonas haloplanctis* A23. J. Biol. Chem. 267: 5217-5221.

Feller, G., Zekhnini, Z., Lamotte-Brasseur, J., and Gerday, C. 1997. Enzymes from cold-adapted microorganisms. The class C beta-lactamase from the Antarctic psychrophile *Psychrobacter immobilis* A5. Eur. J. Biochem. 244: 186-191.

Forterre, P. 1999. Did we have ancestors fond of heat? La Recherche. 317: 36-43.

Francis, K.P., Mayr, R., von Stetten, F., Stewart, G.S., and Scherer, S. 1998. Discrimination of psychrotrophic and mesophilic strains of the *Bacillus cereus* group by PCR targeting of major cold shock protein genes. Appl. Environ. Microbiol. 64: 3525-3529.

Francis, K.P., and Stewart, G.S. 1997. Detection and speciation of bacteria through PCR using universal major cold-shock protein primer oligomers. J. Ind. Microbiol. Biotechnol. 19: 286-293.

Friedman, H., Lu, P., and Rich, A. 1969. An in vivo block in the initiation of protein synthesis. Cold Spring Harb. Symp. Quant. Biol. 34: 255-260.

Fujii, S., Nakasone, K., and Horikoshi, K. 1999. Cloning of two cold shock genes, *cspA* and *cspG*, from the deep-sea psychrophilic bacterium *Shewanella violacea* strain DSS12. FEMS Microbiol. Lett. 178: 123-128.

Galtier, N., Tourasse, N., and Gouy, M. 1999. A nonhyperthermophilic common ancestor to extant life forms. Science. 283: 220-221.

Gerike, U., Danson, M.J., Russell, N.J., and Hough, D.W. 1997. Sequencing and expression of the gene encoding a cold-active citrate synthase from an Antarctic bacterium, strain DS2-3R. Eur. J. Biochem. 248: 49-57.

Goldstein, J., Pollitt, N.S., and Inouye, M. 1990. Major cold shock protein of *Escherichia coli*. Proc. Natl. Acad. Sci. USA. 87: 283-287.

Gounot, A.M. 1991. Bacterial life at low temperature : physiological aspects and biotechnological implications. J. Appl. Bact. 71: 386-397.

Gounot, A.M., and Russell, N.J. 1999. Physiology of cold-adapted microorganisms. In: Cold-Adapted Organisms. Ecology, physiology, enzymology and molecular biology. R. Margesin and F. Schinner, eds. Springer Verlag, Berlin-Heidelberg. p. 3-15.

Graumann, P., and Marahiel, M.A. 1996. Some like it cold: response of microorganisms to cold shock. Arch. Microbiol. 166: 293-300.

Graumann, P., Schroder, K., Schmid, R., and Marahiel, M.A. 1996. Cold shock stress-induced proteins in *Bacillus subtilis*. J. Bacteriol. 178: 4611-4619.

Graumann, P., Wendrich, T.M., Weber, M.H.W., Schroder, K., and Marahiel, M.A. 1997. A family of cold shock proteins in *Bacillus subtilis* is essential for cellular growth and for efficient protein synthesis at optimal and low temperatures. Mol. Microbiol. 25: 741-756.

Graumann, P., and Marahiel, M.A. 1999. Cold shock response in *Bacillus subtilis*. J. Mol. Microbiol. Biotechnol. 1: 203-209.

Gumley, A.W., and Inniss, W.E. 1996. Cold shock proteins and cold acclimation proteins in the psychrotrophic bacterium *Pseudomonas putida* Q5 and its transconjugant. Can. J. Microbiol. 42: 798-803.

Hébraud, M., Dubois, E., Potier, P., and Labadie, J. 1994. Effect of growth temperatures on the protein levels in a psychrotrophic bacterium, *Pseudomonas fragi*. J. Bacteriol. 176: 4017-4024.

Herbert, R. 1986. Psychrophilic organisms. In: Microbes in Extreme Environments. R. Herbert, and G. Codd, eds. Academic Press, London. p. 1-23.

Jones, P.G., and Inouye, M. 1994. The cold-shock response – A hot topic. Mol. Microbiol. 11: 811-818.

Jones, P.G., Krah, R., Tafuri, S.R., and Wolffe, A.P. 1992. DNA gyrase, CS 7.4, and the cold shock response in *Escherichia coli*. J. Bacteriol. 174: 5798-5802.

Jones, P.G., Mitta, M., Kim, Y., Jiang, W., and Inouye, M. 1996. Cold shock induces a major ribosomal-associated protein that unwinds double-stranded RNA in *Escherichia coli*. Proc. Natl. Acad. Sci. USA. 93: 76-80.

Jones, P.G., Van Bogelen, R.A., and Neidhardt, F.C. 1987. Induction of proteins in response to low temperature in *Escherichia coli*. J. Bacteriol. 169:

2092-2095.

Julseth, C.R., and Inniss, W.E. 1990. Induction of protein synthesis in response to cold shock in the psychrotrophic yeast *Trichosporon pullulans*. Can. J. Microbiol. 36: 519-524.

Kim, W.S., and Dunn, N.W. 1997. Identification of a cold shock gene in lactic acid bacteria and the effect of cold shock on cryotolerance. Curr. Microbiol. 35: 59-63.

Kim, W.S., Khunajakr, N., Ren, J., and Dunn, N.W. 1998. Conservation of the major cold shock protein in lactic acid bacteria. Curr. Microbiol. 37: 333-336.

Kim, S.Y., Hwang, K.Y., Kim, S.H., Sung, H.C., Han, Y.S., and Cho, Y. 1999. Structural basis for cold adaptation. Sequence, biochemical properties, and crystal structure of malate dehydrogenase from a psychrophile *Aquaspirillium arcticum*. J. Bio. Chem. 274: 11761-11767.

Kulakova, L., Galkin, A., Kurihara, T., Yoshimura, T., and Esaki, N. 1999. Cold-active serine alkaline protease from the psychrotrophic bacterium *Shewanella* strain ac10: gene cloning and enzyme purification and characterization. Appl. Environ. Microbiol. 65: 611-617.

La Teana, A., Brandi, A., Falconi, M., Spurio, R., Pon, C.L., and Gualerzi, C.O. 1991. Identification of a cold shock transcriptional enhancer of the *Escherichia coli* gene encoding nucleoid protein H-NS. Proc. Natl. Acad. Sci. USA. 88: 10907-10911.

Lelivelt, M., and Kawula, T. 1995. Hsc66, an Hsp70 homolog in *Escherichia coli*, is induced by cold shock but not by heat shock. J. Bacteriol. 177: 4900-4907.

Lottering, E.A., and Streips, U.N. 1995. Induction of cold shock protein in *Bacillus subtilis*. Curr. Microbiol. 30: 193-99.

Mayo, B., Derzelle, S., Fernandez, M., Leonard, C., Ferain, T., Hols, P., Suarez, J.E., and Delcour, J. 1997. Cloning and characterization of *cspL* and *cspP*, two cold-inducible genes from *Lactobacillus plantarum*. J. Bacteriol. 179: 3039-3042.

Mayr, B., Kaplan, T., Lechner, S., and Scherer, S. 1996. Identification and purification of a family of dimeric major cold shock protein homologs from the psychrotrophic *Bacillus cereus* WSBC 10201. J. Bacteriol. 178: 2916-2925.

McElhaney, R.N. 1982. The use of differential scanning calorimetry and differential thermal analysis in studies of model and biological membranes. Chem. Phys. Lipids. 30: 229-259.

McGovern, V., and Oliver, J. 1995. Induction of cold-responsive proteins in *Vibrio vulnificus*. J. Bacteriol. 177: 4131-4133.

Michel, V., Labadie, J., and Hébraud, M. 1996. Effect of different temperature upshifts on protein synthesis by the psychrotrophic bacterium *Pseudomonas fragi*. Curr. Microbiol. 33: 16-25.

Michel, V., Lehoux, I., Depret, G., Anglade, P., Labadie, J., and Hébraud, M. 1997. The cold shock response of the psychrotrophic bacterium *Pseudomonas fragi* involves four low-molecular-mass nucleic acid-binding proteins. J. Bacteriol. 179: 7331-7342.

Morita, R.Y. 1975. Psychrophilic bacteria. Bacteriol. Rev. 39: 144-167.

Neuhaus, K., Francis, K.P., Rapposch, S., Görg, A., and Scherer, S. 1999. Pathogenic *Yersinia* species carry a novel, cold-inducible major cold shock protein tandem gene duplication producing both bicistronic and monocistronic mRNA. J. Bacteriol. 181: 6449-6455.

Ng, H., Ingraham, J.L., and Marr, A.G. 1962. Damage and derepression in *Escherichia coli* resulting from growth at low temperature. J. Bacteriol. 84: 331-339.

Okuyama, H., and Yamada, K. 1979. Specificity and selectivity of diacylglycerolphosphate synthesis in *Escherichia coli*. Biochim. Biophys. Acta. 573: 207-211.

Panoff, J.M., Legrand, S., Thammavongs, B., and Boutibonnes, P. 1994. The cold-shock response in *Lactococcus lactis* subsp. *lactis*. Curr. Microbiol. 29: 213-216.

Perl, D., Welker, C. Schindler, T., Schröder, K., Marahiel, M.A., Jaenicke, R., and Schmid, F.X. 1998. Conservation of rapid two-state folding in mesophilic, thermophilic and hyperthermophilic cold shock proteins. Nat. Struct. Biol. 5: 229-235.

Perrin, C., Guimont, C., Bracquart, P., and Gaillard, J.L. 1999. Expression of a new cold shock protein of 21.5 kDa and of the major cold shock protein by *Streptococcus thermophilus* after cold shock. Curr. Microbiol. 39: 342-347.

Phadtare, S., Alsina, J., and Inouye, M. 1999. Cold shock response and cold shock proteins. Curr. Opin. Microbiol. 2: 175-180.

Phan-Thanh, L., and Gormon, T. 1995. Analysis of heat and cold shock proteins in *Listeria* by two-dimensional electrophoresis. Electrophoresis. 16: 444-450.

Potier, P., Drevet, P., Gounot, A.M., and Hipkiss, A.R. 1987a. ATP-dependent and independant protein degradation in extracts of the psychrotrophic bacterium *Arthrobacter* sp S155. J. Gen. Microbiol. 133: 2797-2806.

Potier, P., Drevet, P., Gounot, A.M., and Hipkiss, A.R. 1987b. Protein turnover in a psychrotrophic bacterium: proteolytic activity in extracts of cells grown at different temperatures. FEMS Microbiol. Lett. 44: 267-271.

Potier, P., Drevet, P., Gounot, A.M., and Hipkiss, A.R. 1990. Temperature-dependant changes in proteolytic activities and protein composition in the psychrotrophic bacterium *Arthrobacter globiformis* S1/55. J. Gen. Microbiol. 136: 283-291.

Ray, M.K., Sitaramamma, T., Ghandi, S., and Shivaji, S. 1994. Occurrence and expression of cspA, a cold shock gene, in Antarctic psychrotrophic bacteria. FEMS Microbiol. Lett. 116: 55-60.

Rentier-Delrue, F., Mande, S.C., Moyens, S., Terpstra, P., Mainfroid, V., Goraj, K., Lion, M., Hol, W.G., and Martial, J.A. 1993. Cloning and overexpression of the triosephosphate isomerase genes from psychrophilic and thermophilic bacteria. Structural comparison of the predicted protein sequences. J. Mol. Biol. 229: 85-93.

Roberts, M.E., and Inniss, W.E. 1992. The synthesis of cold shock proteins and cold acclimation proteins in the psychrophilic bacterium *Aquaspirillum arcticum*. Curr. Microbiol. 25: 275-728.

Russell, N.J. 1989. Adaptive modifications in membranes of halotolerant and halophilic microorganisms. J. Bioenerg. Biomembr. 21: 93-113.

Russell, N.J. 1990. Cold adaptation of microorganisms. Philos. Trans. Roy. Soc. Lond. 326: 595-611.

Russell, N.J., and Fukunaga, N. 1990. A comparison of thermal adaptation of membrane lipids in psychrophilic and thermophilic bacteria. FEMS Microbiol. Rev. 75: 171-182.

Schröder, K., Zuber, U., Willimsky, G., Wagner, B., and Marahiel, M.A. 1993. Mapping of the *Bacillus subtilis cspB* gene and cloning of its homolgs in thermophilic, mesophilic and psychrotrophic bacilli. Gene. 136: 277-280.

Shaw, M.K., and Ingraham, J.L. 1967. Synthesis of macromolecules by *Escherichia coli* near the minimal temperature for growth. J. Bacteriol. 94: 157-164.

Stamm, I., Leclerque, A., and Plaga, W. 1999. Purification of cold shock like proteins from *Stigmatella auriantiaca*: cloning and characterization of the *csp* gene. Arch. Microbiol. 172: 175-181.

Szer, W. 1970. Cell-free protein synthesis at 0°C: an activating factor from ribosomes of a psychrophilic microorganism. Biochim. Biophys. Acta. 213: 159-170.

Thieringer, H.A., Jones, P.G., and Inouye, M. 1998. Cold shock and adaptation. Bioessays. 20: 49-57.

Wang, N., Yamanaka, K., and Inouye, M. 1999. CspI, the ninth member of the CspA family of *Escherichia coli*, is induced upon cold shock. J. Bacteriol. 181: 1603-1609.

Welker, C., Bohm, G., Schurig, H., and Jaenicke, R. 1999. Cloning, overexpression, purification, and physicochemical characterization of a cold shock protein homolog from the hyperthermophilic bacterium *Thermotoga maritima*. Protein Sci. 8: 394-403.

Whyte, L.G., and Inniss, W.E. 1992. Cold shock proteins and cold acclimation proteins in a psychrotrophic bacterium. Can. J. Microbiol. 38: 1281-1285.

Willimsky, G., Bang, H., Fisher, G., and Marahiel, M.A. 1992. Characterization of *cspB*, a *Bacillus subtilis* inductible cold shock gene affecting cell viability at low temperature. J. Bacteriol. 174: 6326-6335.

Wolffe, A.P. 1995. The cold-shock response in bacteria. Sci. Prog. 78: 301-310.

Wouters, J.A., Sanders, J.W., Kok, J., de Vos, W.M., Kuipers, O.P., and Abee, T. 1998. Clustered organization and transcriptional analysis of a family of five *csp* genes of *Lactococcus lactis* MG1363. Microbiol. 144: 2885-2893.

Wouters, J.A., Rombouts, F.M., de Vos, W.M., Kuipers, O.P., and Abee, T. 1999. Cold shock proteins and low-temperature response of *Streptococcus thermophilus* CNRZ302. Appl. Environ. Microbiol. 65: 4436-4442.

Yamanaka, K. 1999. Cold shock response in *Escherichia coli*. J. Mol. Microbiol. Biotechnol. 1: 193-202.

Yamanaka, K., Inouye, M., and Inouye, S. 1999. Identification and characterization of five *cspA* homologous genes from *Myxococcus xanthus*. Biochim. Biophys. Acta. 1447: 357-365.

5

Responses to Cold Shock in Cyanobacteria

Dmitry A. Los[1,2], and Norio Murata[1]

[1]National Institute for Basic Biology, Okazaki, Japan
[2]Institute of Plant Physiology, Russian Academy of Sciences, Moscow, Russia

Abstract

Acclimation of cyanobacteria to low temperatures involves induction of the expression of several families of genes. Fatty acid desaturases are responsible for maintaining the appropriate fluidity of membranes under stress conditions. RNA-binding proteins, which presumably act analogously to members of the bacterial Csp family of RNA chaperones, are involved in the maintenance of the translation under cold stress. The RNA helicase, whose expression is induced specifically by cold, might be responsible for modifying inappropriate secondary structures of RNAs induced by cold. The cold-inducible family of Clp proteins appears to be involved in the proper folding and processing of proteins. Although genes for cold-inducible proteins in cyanobacteria are heterogeneous, some common features of their untranslated regulatory regions suggest the existence of a common factor(s) that might participate in regulation of the expression of these genes under cold-stress conditions. Studies of the patterns of expression of cold-inducible genes in cyanobacteria have revealed the presence of a cold-sensing mechanism that is associated with their membrane lipids. Available information about cold-shock responses in cyanobacteria and molecular mechanisms of cold acclimation are reviewed in this article.

Introduction

Cyanobacteria are unusual prokaryotic microorganisms that have the ability to perform oxygenic photosynthesis (Margulis, 1975) and they might represent some of the most ancient life forms on earth (Schopf et al., 1965). Cyanobacteria can be found all over the world and in environments from

Antarctica, where temperatures never exceed -20°C (Psenner and Sattler, 1998), to hot springs, where temperatures reach 70°C (Ward et al., 1998). Cyanobacteria found in water pockets of Antarctic lake ice, where temperatures are always below 0°C, are metabolically active and capable of performing oxygenic photosynthesis (Prisku et al., 1998). Some species of Synechococcus that are characterized by optimum growth temperatures of 55-60°C are also able to fix CO_2 by photosynthesis at these high temperatures (Meeks and Castenholz, 1971). Thus, the cyanobacteria include psychrophilic, psychrotrophic, mesophilic and thermophilic species that differ from one another with respect to optimal temperatures for growth and the extent of their ability to tolerate temperature stress.

Unicellular and filamentous cyanobacteria have several features that make them particularly suitable for studies of stress responses at the molecular level. The general features of the plasma and thylakoid membranes of cyanobacteria are similar to those of higher-plant chloroplasts in terms of lipid composition and the assembly of proteins. Therefore, cyanobacteria appear to provide a powerful model system for studies of molecular mechanisms of the responses and acclimation of plants to stresses of various kinds (Murata and Wada, 1995; Glatz et al., 1999).

Some strains of cyanobacteria (Synechocystis sp. PCC 6803, and Synechococcus sp. PCC 7942 and PCC 7002) are naturally able to incorporate foreign DNA that is integrated into the genome by high-frequency homologous recombination (Williams, 1988; Haselkorn, 1991). In other strains, such as the filamentous cyanobacterium Anabaena sp. PCC 7120, a method for transformation was developed that is based on the use of plasmids with a broad host-range and bacterial conjugation (Elhai and Wolk, 1988). Since cyanobacteria are characterized by active homologous recombination (Williams and Szalay, 1983; Dolganov and Grossman, 1993), they are widely used in studies of photosynthesis for the production of mutants with disruptions in genes of interest (for review, see Vermaas, 1998).

The complete nucleotide sequence of the genome of Synechocystis sp. PCC 6803 has been determined (Kaneko et al., 1996; Kaneko and Tabata, 1997) and the annotated data is now available via the internet (Nakamura et al., 1998). Furthermore, since random mutagenesis of cyanobacteria can be achieved using transposons (Cohen et al., 1994; Schwartz et al., 1998) or antibiotic-resistance cartridges (Labarre et al., 1989), fully sequenced genes can be randomly disrupted and their functions can be examined under certain stress conditions. The complete sequence of the genome facilitates the localization of the sites of mutations and the identification of relevant genes.

Responses of cyanobacterial cells to cold stress are basically of two types. One type involves the cold-induced desaturation of fatty acids in membrane lipids, such that the membranes become less rigid to compensate for the decrease in membrane fluidity that would otherwise occur at the low temperature (Murata and Los, 1997). The other type involves the cold-induced synthesis of enzymes that enhance the efficiency of transcription and translation to compensate for the decrease in the efficiency of these processes that would otherwise occur at the low temperature (Sato, 1994, 1995). Both types of response serve to protect the cyanobacteria from the detrimental effects of cold stress.

Cold-Inducible Genes and their Regulation in Cyanobacteria

The families of cold-inducible genes in cyanobacteria that have been reported to date are listed in Table 1. The first cold-inducible genes to be characterized in cyanobacteria were genes for fatty acid desaturases (Wada et al., 1990; Murata and Wada, 1995). The fatty acid desaturases maintain the appropriate physical state of the cell membranes (Murata and Los, 1997). Subsequently, genes for RNA-binding proteins (Rbps) were identified as cold-inducible genes (Sato, 1995). The Rbps appear to act similarly to the Csp RNA chaperones of *Escherichia coli* and *Bacillus subtilis*. The S21 protein in the small subunit of ribosomes was also shown to be induced by cold and to accumulate transiently in ribosomes at low temperatures (Sato, 1994). Next, Clp proteins were discovered as a novel family of cold-shock chaperones and proteases (Celerin et al., 1998). Most recently, genes for RNA helicases (Chamont et al., 1999), which appear to establish the appropriate secondary structure of mRNAs, were identified as being inducible by cold.

Desaturases
Acyl-Lipid Desaturases
There are three types of fatty acid desaturase and the acyl-lipid desaturases are one such type (Murata and Wada, 1995). They are characterized by their ability to convert a single bond in a fatty acyl chain, which has been esterified to a membrane glycerolipid, into a double bond. In other words, they catalyze the introduction of individual double bonds (Sato et al., 1979; Sato and Murata, 1980; Murata and Wada, 1995). The desaturation of fatty acids and the expression of genes for desaturase have been extensively studied in *Synechocystis* sp. PCC 6803. This cyanobacterium has four acyl-lipid desaturases (Figure 1A), which catalyze desaturation at the $\Delta 9$, $\Delta 12$, $\Delta 6$ and $\omega 3$ positions, respectively, of fatty acids that are located at the *sn*-1 position of the glycerol moieties of glycerolipids (Murata et al., 1992; Higashi and Murata, 1993). The $\Delta 9$ desaturase introduces the first unsaturated bond into stearic acid to produce oleic acid, which is further desaturated to linoleic acid by the $\Delta 12$ desaturase. The $\Delta 6$ and $\omega 3$ desaturases then introduce unsaturated bonds to generate tri- and tetraunsaturated fatty acids (Figure 1A). By contrast to *Synechocystis* sp. PCC 6803, *Synechococcus* sp. PCC 7942 has only one gene for a desaturase, namely, the *desC* gene for the $\Delta 9$ desaturase. Thus, the cells contain saturated and monounsaturated fatty acids but no polyunsaturated fatty acids (Figure 1B).

Expression of the Genes for Fatty Acid Desaturases in Synechocystis sp. PCC 6803
The *desA* gene for the $\Delta 12$ desaturase in *Synechocystis* sp. PCC 6803 was the first gene for an acyl-lipid desaturase to be cloned (Wada et al., 1990),

Table 1. Genes Known to be Induced by Cold Stress in Cyanobacteria

Gene	Gene product	Cyanobacterium	Reference
Desaturase family			
desA	Δ12 desaturase	Synechocystis sp. PCC 6803 Synechocystis sp. PCC 6714 Synechococcus sp. PCC 7002 Spirulina platensis	Wada et al., 1990 Sakamoto et al., 1994a Sakamoto and Bryant, 1997 Murata et al., 1996
desB	ω3 desaturase	Synechocystis sp. PCC 6803 Synechococcus sp. PCC 7002	Sakamoto et al., 1994c Sakamoto and Bryant, 1997
desC	Δ9 desaturase	Synechococcus sp. PCC 7942 Synechococcus sp. PCC 7002	Ishizaki-Nishizawa et al., 1996 Sakamoto and Bryant, 1997
desD	Δ6 desaturase	Synechocystis sp. PCC 6803 Spirulina platensis	Reddy et al., 1994 Murata et al., 1996
Rbp family			
rbpA1 rbpA2 rbpA3 rbpC	RNA-binding protein (RbpA1) RNA-binding protein (RbpA2) RNA-binding protein (RbpA3) RNA-binding protein (RbpC)	Anabaena variabilis M3 Anabaena variabilis M3 Anabaena variabilis M3 Anabaena variabilis M3	Sato and Nakamura, 1998 Sato, 1995 Sato and Maruyama, 1997 Sato, 1995
RNA helicases			
crhB crhC	RNA helicase (CrhB) RNA helicase (CrhC)	Anabaena sp. PCC 7120 Anabaena sp. PCC 7120	Chamot et al., 1999 Chamot et al., 1999
Clp family			
clpB clpP1 clpX	Molecular chaperone (ClpB) Protease (ClpP) Unknown	Synechococcus sp. PCC 7942 Synechococcus sp. PCC 7942 Synechococcus sp. PCC 7942	Porankiewicz and Clarke, 1997 Porankiewicz et al., 1998 Porankiewicz et al., 1998
Others			
rpsU lti2	Ribosomal subunit (S21) Unknown (Lti2)	Anabaena variabilis M3 Anabaena variabilis M3	Sato, 1994 Sato, 1992

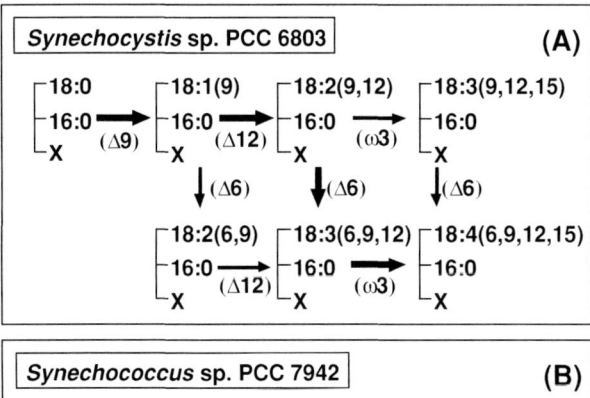

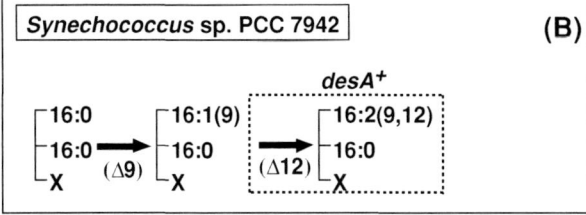

Figure 1. Schematic Representation of the Desaturation of Fatty Acids in the Membrane Lipids of two Species of Cyanobacteria. Glycerolipids esterified exclusively with saturated fatty acids are synthesized first. Then the fatty acids are desaturated to yield their monounsaturated derivatives by acyl-lipid $\Delta 9$ desaturase and these derivatives are converted to polyunsaturated fatty acids by other acyl-lipid desaturases (Sato and Murata, 1980; Higashi and Murata, 1993; Murata and Wada, 1995). (A) Synechocystis sp. PCC 6803. Thick and thin arrows indicate major and minor pathways, respectively. (B) Synechococcus sp. PCC 7942. The reaction enclosed by a dotted line has been introduced by transformation with the desA gene for the $\Delta 12$ desaturase from Synechocystis sp. PCC 6803. X represents the polar head group. Adapted from Murata et al. (1992).

and the low temperature-dependent expression of this gene has been studied extensively (Los et al., 1993; 1997). The level of the transcript increases 10-fold within 30 min of a downward shift in temperature from 34°C to 22°C. The accumulation of the transcript is caused by both the activation of transcription and the enhanced stabilization of the mRNA at low temperature (Los and Murata, 1994; Los et al., 1997). When the temperature returns to normal, the desA mRNA rapidly disappears (Los et al., 1997).

The desD gene for the $\Delta 6$ desaturase is also induced by a downward shift in temperature, as demonstrated by Northern and Western blotting analyses (Los et al., 1997). The level of desD mRNA increases about 10-fold within in 15 min, while the level of the enzyme itself doubles within 4 h of the start of cold treatment.

Of the three cold-inducible genes for desaturases in Synechocystis sp. PCC 6803, it is the desB gene for the $\omega 3$ desaturase that responds most dramatically to a decrease in temperature: The level of desB mRNA increases 15-fold within 10 min after a shift in temperature from 34°C to 22°C (Los et al., 1997). The accumulation of desB mRNA due to both the acceleration of transcription and the stabilization of the transcript. Figure 2 shows the kinetics of accumulation of desB mRNA, of the $\omega 3$ desaturase, and of $\omega 3$-unsaturated fatty acids [α-linolenic (α-18:3) and stearidonic (18:4) acids] in Synechocystis

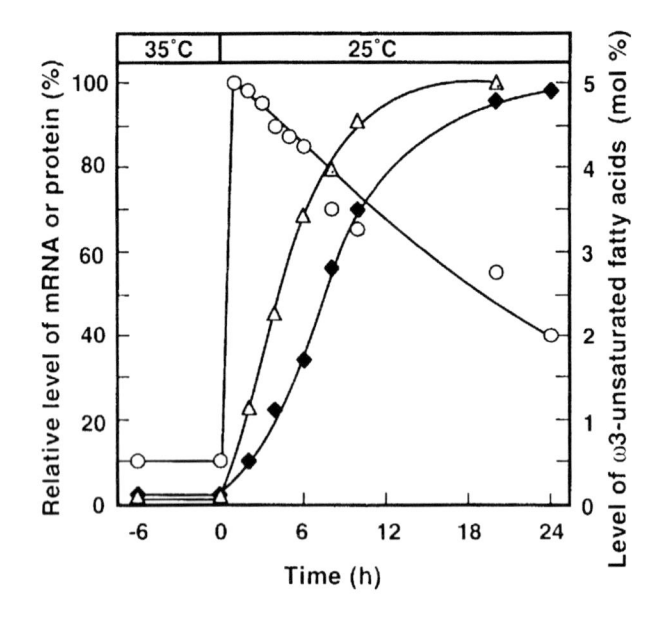

Figure 2. Changes in *Synechocystis* sp. PCC 6803 in levels of the transcript of *desB*, the encoded ω3 desaturase and ω3-unsaturated fatty acids after a shift in temperature from 35°C to 25°C. O-O, *desB* mRNA; Δ-Δ, ω3 desaturase; ■-■, ω3-unsaturated fatty acids (*i.e.*, the level of α-18:3 plus 18:4 relative to the total fatty acids).

sp. PCC 6803 after a shift in temperature from 35°C to 25°C. As shown in Figure 2, *desB* mRNA accumulates rapidly and then its level starts to decline gradually. The ω3 desaturase is barely detectable at 35°C but its level becomes gradually higher as time passes after the shift in temperature. Then, the level of the ω3 desaturase remains high for 10 h (Figure 2). The accumulation of the ω3 desaturase is followed by the slow and gradual accumulation of ω3-unsaturated fatty acids.

Primer extension analysis indicated that transcription of the *desA*, *desB*, and *desD* genes starts at positions −114, -35, and −347, respectively, counted from the site of transcription start, at both 34°C and 22°C. Thus, it appears that RNA polymerase utilizes the same promoters at both temperatures. An alignment of nucleotide sequences near the sites of initiation of transcription of the genes for the cold-inducible desaturases revealed a consensus sequence, GTTTGTTTT, just downstream of these sites, irrespective of the position of each initiation site (see also Section 4).

Although expression of genes for desaturases has been studied most extensively in *Synechocystis* sp. PCC 6803, there are several reports of the cold-induced expression of genes for desaturases in other species of cyanobacteria (Table 1). The *desC* gene in *Synechococcus* sp. PCC 7002 is cold-inducible and the level of its transcript increases markedly within 15 min of a shift from 38°C to 22°C (Sakamoto and Bryant, 1997). The *desA* and *desB* genes are also induced rapidly in this cyanobacterium (Sakamoto et al., 1997). In *Synechococcus* sp. PCC 7942, which only has a *desC* gene for the Δ9 desaturase (Figure 1B), the gene is induced 30 min after a shift from 36°C to 24°C (Ishizaki-Nishizawa et al., 1996).

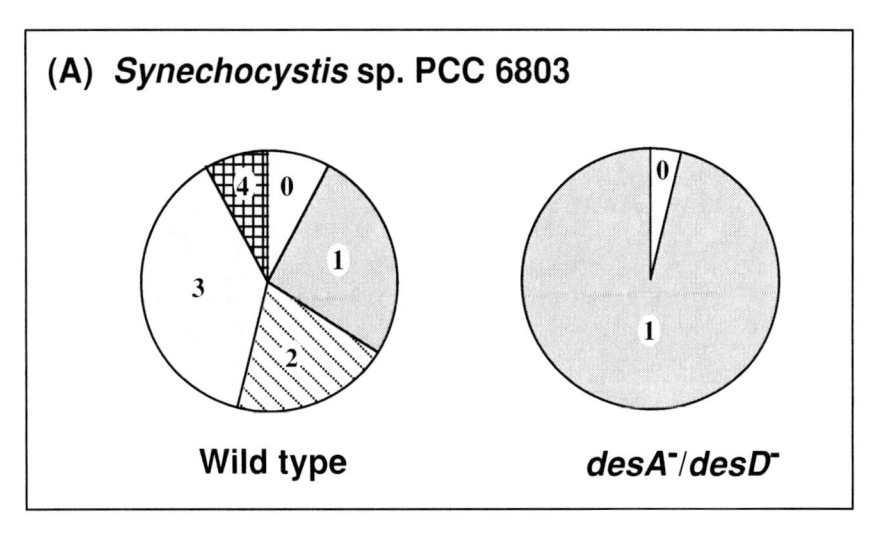

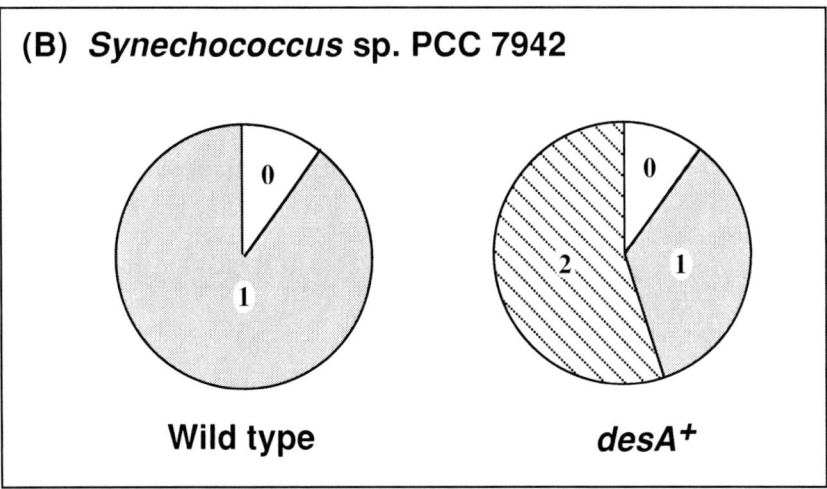

Figure 3. Changes in the Unsaturation of Fatty Acids in Cyanobacterial Cells by Genetic Manipulation of Acyl-Lipid Desaturases. (A) The fatty acid composition of glycerolipids in wild-type and $desA^-/desD^-$ cells of Synechocystis sp. PCC 6803 after growth at 25°C. (B) The fatty acid composition of glycerolipids in wild-type and $desA^+$ cells of Synechococcus sp. PCC 7942 after growth at 25°C. Numbers in the pie charts represent the numbers of double bonds in the individual molecular species of lipids. Adapted from Tasaka et al. (1996) and Los and Murata (1998).

Biological Functions of Fatty Acid Desaturases

The importance of the desaturases and of the expression of their genes in the acclimation to cold of cyanobacteria has been well documented (Wada et al., 1990; Gombos et al., 1992; 1994; Wada et al., 1994; Tasaka et al., 1996; for reviews, see Murata and Wada, 1995; Los and Murata, 1998). A series of mutants of Synechocystis sp. PCC 6803, which are defective in a stepwise manner in the desaturation of fatty acids in membrane lipids, was generated by targeted mutagenesis of individual desaturases (Tasaka et al., 1996). Targeted mutagenesis of both the *desA* gene for the Δ12 desaturase and *desD* gene for the Δ6 desaturase resulted in dramatic changes in the fatty acids in membrane lipids. There was a considerable increase in the level of monounsaturated oleic acid at the expense of polyunsaturated fatty acids, such as di-, tri-, and tetraunsaturated fatty acids (Figure 3A). Cells with these two mutations grew as well as wild-type cells at 35°C but did not grow well at 25°C. At 20°C, the *desA⁻/desD⁻* mutant cells were unable to propagate, whereas wild-type cells grew relatively well at this temperature (Tasaka et al., 1996). Moreover, *desA⁻/desD⁻* cells failed to recover from photo-induced damage to the photosystem II complex at low temperatures. Apparently, they were unable to process the precursor to the D1 protein to generate the mature D1 protein, an essential component of the reaction center of the photosystem II complex (Kanervo et al., 1997). Since only two genes for fatty acid desaturases were inactivated in the *desA⁻/desD⁻* strain, we can reasonably conclude that the ability of *Synechocystis* sp. PCC 6803 to tolerate cold stress is determined by the presence of polyunsaturated fatty acids.

A similar conclusion is reached with another set of experiments that involved transformation of *Synechococcus* sp. PCC 7942, which normally contains only saturated and monounsaturated fatty acids (Figure 1B), with the *desA* gene for the Δ12 desaturase of *Synechocystis* sp. PCC 6803 (Figure 3B). The transformed *desA⁺* cells synthesized diunsaturated fatty acids at the expense of monounsaturated fatty acids, and they were able to tolerate lower temperatures than wild-type cells (Wada et al., 1990; 1994). Moreover, the *desA⁺* cells appeared to be more tolerant to photoinhibition (Gombos et al., 1997). During cold acclimation, wild-type cells of *Synechococcus* sp. PCC 7942 replace one isoform of the D1 protein (D1:1) with another isoform (D1:2) within a few hours (Campbell et al., 1995). This replacement appears to be essential for the survival of this cyanobacterium at low temperatures. The transformation with the *desA* gene shifted the temperature critical for the replacement of the D1:1 form with the D1:2 form toward a lower temperature (Sippola et al., 1998). Although the molecular mechanism for the shift in the critical temperature is not known, these observations reveal that the desaturation of fatty acids in membrane lipids is an important factor in the acclimation to cold.

The Rbp Family

Rbps (RNA binding proteins) are involved in various aspects of the metabolism of RNA, such as splicing, modification, maintenance of stability and translation (Kenan et al., 1991; Nagai et al., 1995). The Rbps in the chloroplasts of higher plants contain two RNA-recognition motifs, an amino-terminal acidic domain and a carboxy-terminal glycine-rich domain (Ye et al., 1991). Sato (1995) characterized the family of cold-inducible genes that encode Rbps in *Anabaena variabilis* strain M3. The Rbps of this cyanobacterium can be divided into those with a glycine-rich carboxy-terminal domain and those without such a domain.

Sato (1995) demonstrated that, among eight *rbp* genes found in

Anabaena variabilis, four genes (*rbpA1*, *rbpA2*, *rbpA3*, *rbpB* and *rbpC*) for proteins, that contain glycine-rich domains are regulated by cold. Two genes (*rbpB* and *rbpD*) for proteins without glycine-rich domains are expressed more or less constitutively (Sato, 1995; Sato and Maruyama, 1997). The mRNAs for *rbpA1* and *rbpA2* are barely detectable at 38°C. *RbpA1* mRNA becomes easily detectable 30 min after cells have been transferred from 38°C to 22°C. The level of this mRNA reaches a maximum within 3 h and then gradually decreases (Sato, 1995). Expression of *rbpA2* is induced within 10 min after transfer from 38°C to 22°C. The level of its transcript reaches a maximum in 2 h, and then declines gradually. *RbpC* mRNA is detectable at 38°C, and its level increases about 10-fold within 1 h after transfer from 38°C to 22°C (Sato, 1995). The levels of the corresponding proteins also increase dramatically after cold shock. However, whereas levels of transcripts of the cold-inducible *rbp* genes increase transiently, the levels of the corresponding proteins increase gradually and remain maximal for 24 h after the shift in temperature from 38°C to 22°C (Sato, 1995).

Sato and Maruyama (1997) demonstrated that transcription of the *rbpA3* gene is driven by two different promoters. One of the promoters is active at high temperatures and its activity is suppressed at low temperatures. By contrast, the activity of the other promoter increases transiently after a shift in temperature from 38°C to 22°C. Using a *lacZ* reporter fused to a number of modified promoter regions, Sato and Nakamura identified a putative cold-responsive *cis*-acting element in a 5'-untranslated region (5'-UTR) of the *rbpA1* gene (Sato and Nakamura, 1998). A 150-bp region of DNA, which is located between the site of initiation of transcription and a ribosome-binding site, is absolutely necessary for the cold-induced transcription of the *rbpA1* gene. Deletions within this region result in constitutive transcription at both 38°C and 22°C (Sato and Nakamura, 1998). This observation suggests that transcription of the *rpbA1* gene might be repressed at high temperatures by an unidentified repressor protein. The argument in favor of the existence of such a protein is strengthened by the results of gel mobility shift assays with the target DNA sequence as a probe and a crude extract of proteins from cells grown at a high temperature. A protein(s) that bound to the 5'-UTR of the *rbpA1* gene was present in the crude extract of cells grown at 38°C but not of cells grown at 22°C. Affinity purification of DNA-binding proteins demonstrated the presence of two putative repressor proteins of about 75 kDa and about 32 kDa, respectively (Sato and Nakamura, 1998).

The discovery of cyanobacterial Rbps whose synthesis is regulated by temperature revealed the existence of a new class of stress-inducible RNA-binding proteins. The sequence of the genome of *Synechocystis* sp. PCC 6803 indicates that cyanobacteria do not have the cold-shock proteins (Csps) that are found in some eubacteria (Jones and Inoue, 1994). The synthesis of Csps is rapidly induced by cold and Csps are thought to play important roles in the regulation of transcription under cold stress in *E. coli* and *B. subtilis* (Jones and Inoue, 1994; Schnuchel *et al.*, 1993). Some Csps, known as RNA chaperones, bind to single-stranded DNA and RNA *in vitro* with a molecular surface that corresponds to consisting of a β-sheet, a phenomenon that is analogous to the binding of RNA-recognition motifs to RNAs (Thieringer et al., 1998). Both the Rbps in cyanobacteria and the Csps in E. coli and B. subtilis are induced by cold and bind RNA. Therefore, it is possible that Rbps might function similarly to the Csps.

Cold-Induced RNA Helicases

RNA helicases are responsible for modifying the secondary structure of mRNAs, which is a critical factor in the regulation of translation (Fuller-Pace, 1994). RNA helicases in E. coli play important roles in the assembly of ribosomes (Nishi et al., 1988; Nicol and Fuller-Pace, 1995), the turnover of RNA (Py et al., 1996), and the acclimation to cold (Jones et al., 1996). Two genes for RNA helicases, crhB and crhC, have been identified in Anabaena sp. PCC 7120 (Chamot et al., 1999). The crhB gene is expressed under variety of stress conditions (e.g., cold stress, salt stress, nitrogen limitation), while expression of the crhC gene occurs exclusively in response to cold stress. The crhC gene is expressed specifically in cells that have been transferred from 30°C to 20°C, with the level of crhC mRNA increasing more than 100-fold during incubation of cells at 20°C for 3 h (Chamot et al., 1999).

The deduced amino acid sequence of the CrhC protein led to its identification as a novel RNA helicase that belongs to the DEAD (Asp-Glu-Ala-Asp) box family of helicases (Gorbalenya and Koonin, 1993). However, CrhC has a novel FAT (Phe-Ala-Thr) box instead of the canonical SAT (Ser-Ala-Thr) box that is characteristic of known DEAD box RNA helicases (Py et al., 1996). The suggested role of CrhC in cold acclimation is the destabilization of the secondary structures of mRNAs, which allows cells to overcome the cold-induced blockage of the initiation of translation that occurs at low temperatures (Chamot et al., 1999).

Protein S21, a Component of the Small Subunit of Ribosomes

Cyanobacterial ribosomes are of the prokaryotic type and are similar to those in E. coli (Gray and Herson, 1976; Sato et al., 1998). The rpsU gene for protein S21, a component of the small subunit of ribosomes, is located just downstream of the rbpA1 gene in the genome of Anabaena variabilis strain M3 (Sato, 1994). Two transcripts are characteristic of this gene cluster. The level of the combined transcripts of the rbpA1 and rpsU genes increases 10-fold within 2.5 h of a shift from 38°C to 22°C. By contrast, the minor monocistronic transcript of rpsU is more abundant at 38°C (Sato, 1994). The level of the S21 protein increases 3-fold after the shift in temperature. Sato et al. (1997) demonstrated that, in isolated ribosomes, the S21 protein is present at an equimolar level relative to other ribosomal proteins at 22°C, but the relative level of S21 decreases at high temperatures.

A cold-induced increase in the level of S21 protein has also been found in Synechocystis sp. PCC 6803, in which the rpsU gene is not adjacent to the rbpA gene but is located downstream of the rRNA operon (Sato et al., 1997).

The changes in level of S21 protein in cyanobacterial ribosomes with changes in temperature raise an interesting question about the role of this protein in the acclimation of the translational apparatus to cold stress. It is possible that a ribosome without S21 protein might be inactive under cold conditions and that it is only when S21 is present that the ribosome becomes translationally active. The pattern of the cold-induced accumulation of S21 in cyanobacteria suggests that this protein might be involved in the acclimation to cold of the translational apparatus, whose activity appears always to decrease in prokaryotes upon exposure to cold shock.

In *E. coli*, S21 is a constitutive component of the ribosome. It is likely that nonphotosynthetic bacteria express the *rpsU* gene constitutively and that S21 is always present in ribosomes at an appropriate stoichiometric level (Held *et al.*, 1974). However, the *rpsU* gene is absent from the genome of the parasitic prokaryote *Mycoplasma genitalium* (Frazer *et al.*, 1995) and S21 is absent from chloroplast ribosomes (Harris *et al.*, 1994), as well as from cytoplasmic ribosomes (Sato *et al.*, 1997), in higher plants.

The Clp Family

Caseinolytic proteases (Clps) represent a new family of bacterial molecular chaperones that includes proteases that are expressed constitutively in some cases and stress-inducibly in others (Schrimer *et al.*, 1996; Thompson and Maurizi, 1994; Kessel *et al.*, 1995). The sequence of the genome of *Synechocystis* sp. PCC 6803 (Kaneko *et al.*, 1996) indicates that it contains genes for ClpB, ClpC, ClpP, and ClpX. Moreover, up to four isozymes of ClpP are encoded by a multigene sub-family.

In *Synechococcus* sp. PCC 7942, the *clpP1* gene is found within the *clpP1/clpX* operon (Porankiewicz *et al.*, 1998). Rapid accumulation of ClpP1 is observed under cold stress and also under UV-B light, and the amount of ClpP1 increases 15-fold within 24 h of the start of cold treatment (Porankiewicz *et al.*, 1998).

Growth of a *clpP1* null mutant, Δ*clpP1*, is severely inhibited at low temperatures. Wild-type cells survive and propagate after a shift from 37°C to 25°C, while Δ*clpP1* mutant cells completely lose viability at 25°C (Porankiewicz and Clarke, 1997). During cold acclimation, wild-type cells replace one isoform of the D1 protein (D1:1) with another isoform (D1:2) within just a few hours. Once acclimated to low temperatures, wild-type cells then replace the D1:2 isoform by the D1:1 isoform. By contrast, Δ*clpP1* cells fail to perform this final step (Porankiewicz *et al.*, 1998). Although the mechanism responsible for the exchange of the D1:1 and D1:2 isoforms in this cyanobacterium is poorly understood, such observations demonstrate that ClpP1 is indispensable to the acclimation to cold. It is of particular interest in this context that the ClpP protein exhibits peptidase activity in *E. coli* (Maurizi *et al.*, 1990a,b). ClpP1 in *Synechococcus* sp. PCC 7942 might participate in the exchange of isoforms of the D1 protein by degrading the inappropriate isoform of this protein.

ClpB (HSP100) in *Synechococcus* sp. PCC 7942 was defined initially as a heat-inducible molecular chaperone that is essential for the acquisition of thermotolerance (Eriksson and Clarke, 1996). However, synthesis of ClpB is also strongly induced under cold stress (Porankiewicz and Clarke, 1997;

	Gene	Source	Reference
TGGCAACGTGTTATAAAAAGAAAAGTTTG-TTTACCTG	desA	6803	(Los et al., 1997)
GCCTTCTTTAGGATAGAATCATAGGATTG-TTTTGCCG	desB	6803	(Los et al., 1997)
TAGCAAAATAAGTTTAATTCATAACTGAG-TTTTACtG	desD	6803	(Los et al., 1997)
TCCGAAATTTACATCTCTAGACAGTAACAATTTTG	rbpA1	A.v.	(Sato and Nakamura, 1998)
TCCGAAACCTAAATCTCTACGTACCTATGATTTCG	rbpA2	A.v.	(Sato and Nakamura, 1998)
TCCGAAATTTAAATCTCTACACATTTATGATTTTG	rbpA3	A.v.	(Sato and Maruyama, 1997)
TCCGAAATCTCAATCCCTAGACACTTCTGATTTTG	rbpB	A.v.	(Sato and Nakamura, 1998)
CTTACCATTATGAGCCATTAATTAAGCTAATTTAGCAG	rpsU	A.v.	(Sato et al., 1997)
GCTTAATACTAGCATTTTATATTTTACTGATTTT	lti2	A.v.	(Sato, 1992)
TCCnAAATTTAPATnnATAnATAnnTnTGATTTTGCnG	CONSENSUS		

Figure 4. Alignment of the 5'-Untranslated Regions of Various Cold-Inducible Genes from Cyanobacteria. The sites of initiation of transcription of the genes for desaturases are double-underlined. Part of the 5'-UTR of the rbpA1 gene to which a trans-acting protein factor(s) binds is underlined. See text for further details. Abbreviations: A.v., Anabaena variabilis strain M3; 6803, Synechocystis sp. PCC 6803.

Celerin et al., 1998). Targeted mutagenesis of the clpB gene accelerated inhibition of the activity of the photosystem II complex at low temperatures and reduced the ability of mutant cells to acclimate to low temperatures in terms of propagation and survival. Porankiewicz and Clarke (1997) suggested that ClpB might renature and solubilize aggregated proteins at low temperatures at which translation is repressed.

Expression of the lti2 Gene is Induced by Low Temperature

A low temperature-inducible gene, lti2, was cloned from the genome of Anabaena variabilis strain M3 (Sato, 1992). This gene exhibits significant homology to genes for various α-amylases and to genes for glucanotransferases of bacteria, fungi, and higher plants. The Lti2 protein expressed in vitro has no α-amylase activity (Sato, 1992). Sato suggested that the Lti2 protein might be a kind of glucanotransferase since it contains regions of conserved amino acids that are characteristic of the catalytic centers of glucanotransferases (Sato, 1992).

The level of the transcript of the lti2 gene increases 40-fold within one hour of a shift from 38°C to 22°C. Moreover, a homolog of the lti2 gene in Anabaena sp. PCC 7120 responds to high-salt stress and to osmotic stress in addition to low-temperature stress (Schwartz et al., 1998). An orrA (osmotic response regulator) gene for the regulator of the response of the lti2 gene to salt stress has been identified in Anabaena sp. PCC 7120. The orrA gene is homologous to a regulator of the expression of a gene for an extracellular proteinase in B. subtilis (Kunst et al., 1997). Thus, it has been suggested that a two-component regulatory system exists to control expression of the lti2 gene in response to changes in osmotic pressure (Schwartz et al., 1998). The existence of a cold-response regulator of expression of the lti2 gene and the biological role of the Lti2 protein under stress conditions remain to be established.

Some Common Features of the Regulatory Regions of Cold-Inducible Genes

An alignment of the 5'-untranslated regions (5'-UTRs) of various cyanobacterial genes whose expression is induced by cold reveals some common sequence (Figure 4). The consensus sequence corresponds to part of the 5'-UTR of the *rbpA1* gene of *A. variabilis* to which a *trans*-acting protein factor(s) binds (Sato and Nakamura, 1998). Expression of the *rbpA1* gene is suppressed at high temperatures by this interaction (Sato and Nakamura, 1998). Expression of the *desB* gene for the ω3 desaturase in *Synechocystis* sp. PCC 6803 seems to be suppressed at high temperatures by the interaction of a large acidic protein with the region that includes the consensus sequence (our unpublished results). It remains to be determined whether the putative repressors of all the cold-inducible genes are identical or even similar.

Membrane Fluidity as a Link Between the Temperature of the Environment and the Induction of Gene Expression

The expression of the various above-mentioned genes is induced at low temperatures or after a downward shift in temperature but it remains unclear how cyanobacterial cells detect the ambient temperature or a change in the ambient temperature that leads to the cold-induced expression of genes. The most extensive study of this problem to date has involved expression of the genes for desaturases and its relationship to the dynamics of membrane structure in *Synechocystis* sp. PCC 6803. It is important in this context to recall here that the extent of unsaturation of the fatty acids in membrane lipids is the major factor that determines the fluidity of the membrane (Kates *et al.*, 1984; Cossins, 1994). Several lines of evidence exist for the contribution of either membrane fluidity or the unsaturation of fatty acids in membrane lipids to the perception of the temperature signal, as follows.

1) We compared the temperature-dependent expression of the cold-inducible *desA* gene in two types of cell that had been grown at different temperatures. Fatty acids in cells grown at 32°C were more unsaturated than those in cells grown at 36°C, an indication that membranes in cells grown 32°C were more fluid than those in cells grown at 36°C. In cells grown at 36°C, the cold-induced expression of the *desA* gene began to appear when the temperature was lowered to 28°C, whereas in cells grown at 32°C it began to appear at 26°C (Los *et al.*, 1993). These observations indicated that, in these cyanobacterial cells, an increase in unsaturation of fatty acids in membrane lipids and, thus, in the fluidity of membranes, shifted the cold-induced expression of the *desA* gene toward a lower temperature. It was clear that cyanobacterial cells perceived a change in temperature and not the absolute temperature and, moreover, that the cells sensed a change in temperature only when it exceeded 6°C. In terms of the physical response, this biological perception of temperature is very sensitive since a decrease in temperature of 6°C corresponds to a reduction in the molecular motion in membrane lipids of only 2% (Murata and Los, 1997).

2) We have been able to decrease the fluidity of membranes by the

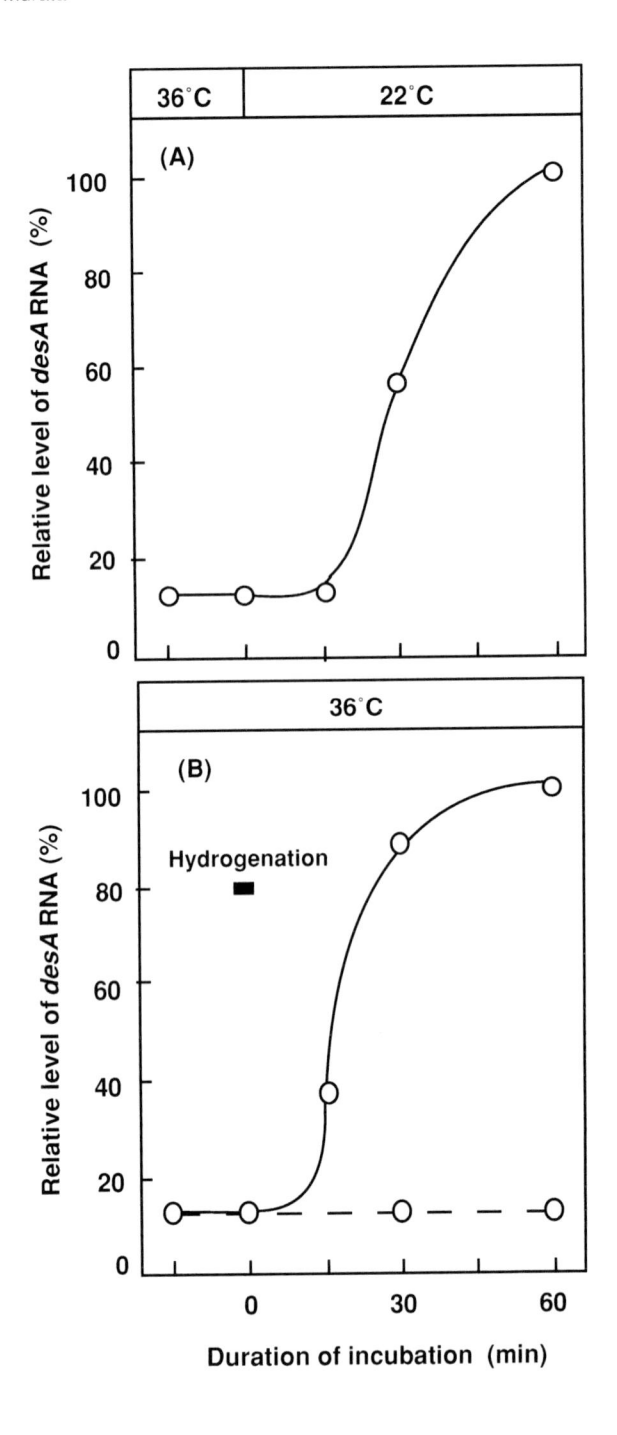

Figure 5. Changes in the Level of the *desA* Transcript in *Synechocystis* sp. PCC 6803. (A) Temperature-induced accumulation of *desA* mRNA. Cells that had been grown at 36°C were transferred to 22°C at time zero. (B) Hydrogenation-induced accumulation of *desA* mRNA. Cells that had been grown at 36°C were subjected to chemical hydrogenation at 36°C for 4 min (indicated by a horizontal bar). Under these conditions, 5% of the fatty acids in the glycerolipids in the plasma membrane were hydrogenated while practically no lipids in the thylakoid membranes were hydrogenated. Adapted from Los et al. (1993) and Vigh et al. (1993).

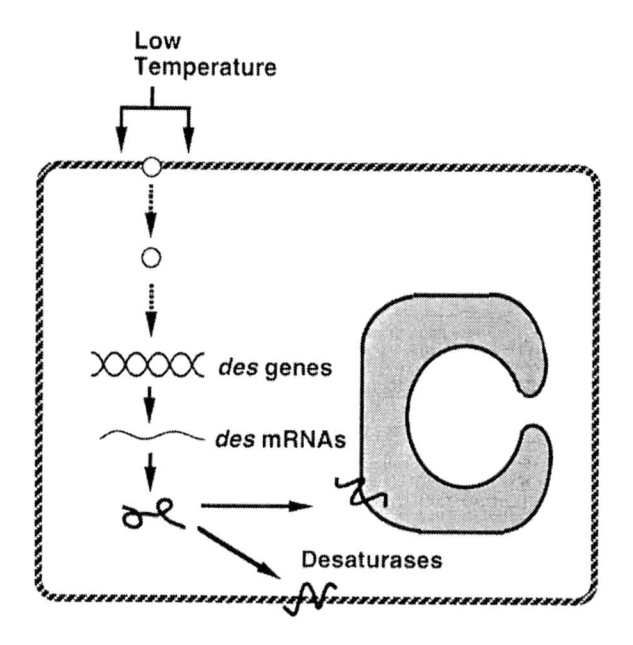

Figure 6. Model of the Induction of Desaturases in Cyanobacterial Cells Proposed model of the perception of temperature and transduction of the signal during the low temperature-induced expression of genes for desaturases in cyanobacterial cells. Putative sensor and signal-transducing components are indicated by circles and dotted arrows, respectively. PM, Plasma membrane; TM, thylakoid membrane; des, desaturase; Pre, precursor to desaturase.

catalytic hydrogenation, under isothermal conditions, of the fatty acids of the lipids in the plasma membrane of *Synechocystis* sp. PCC 6803 (Vigh et al., 1993). This chemical method allowed us to examine the direct effects of membrane fluidity in the presence of only minimal contributions by other changes caused by a change in temperature. Hydrogenation of cells at 36°C for 4 min converted 5% of the unsaturated fatty acids to saturated fatty acids in the glycerolipids of plasma membranes, but not of thylakoid membranes. The decrease in membrane fluidity, caused either by cold stress or by catalytic hydrogenation, resulted in rapid activation of the expression of the *desA* gene (Figure 5). Moreover, hydrogenation of lipids in plasma membranes increased the threshold temperature for the expression of the *desA* gene from 28°C to 30°C (Vigh et al., 1993). These findings suggest that the primary signal in the perception of temperature is a change in the fluidity of the plasma membrane.

3) We compared the temperature-dependent expression of the cold-inducible *desB* gene in wild-type and *desA⁻/desD⁻* cells of *Synechocystis* sp. PCC 6803 (Figure 3A). In wild-type cells that had been grown at 36°C and contained mono-, di-, and triunsaturated fatty acids, transcription of the *desB* gene was induced at 28°C. By contrast, in *desA⁻/desD⁻* cells that had been grown at 36°C and contained monounsaturated but no polyunsaturated fatty acids, transcription of the *desB* gene was induced at 32°C (our unpublished results). These results suggest that the replacement of polyunsaturated by

monounsaturated fatty acids and, therefore, a decrease in membrane fluidity, shifted the cold-induced expression of the *desB* gene toward lower temperatures by 5°C in *desA⁻/desD⁻* cells.

4) We also compared the temperature-dependent expression of *desC*, the cold-inducible gene for the Δ9 desaturase, in wild-type and *desA⁺* cells of *Synechococcus* sp. PCC 7942 (Figure 3B). In wild-type cells that produced monounsaturated but no polyunsaturated fatty acids, induction of expression of the *desC* gene occurred at 30°C. In *desA⁺* cells that produced diunsaturated fatty acids at the expense of monounsaturated fatty acids, induction of the expression of the *desC* gene occurred at 28°C (our unpublished results). These observations indicated that, in *desA⁺* cells, replacement of most of the monounsaturated fatty acids by diunsaturated fatty acids and, therefore, an increase in membrane fluidity shifted the cold-induced expression of the *desC* gene toward higher temperatures by 2°C.

The discovery of an apparent feedback link between membrane fluidity and expression of genes for desaturases suggests that a temperature sensor might be located in the cyanobacterial plasma membrane that perceives changes in membrane fluidity and transmits the signal to activate the expression of the genes for desaturases (Murata and Wada, 1995; Murata and Los, 1997). The putative chain of events from cold shock to the induction of the genes for desaturases is summarized schematically in Figure 6. The pathway from the perception of temperature to the induction of the genes for desaturases is now being investigated in our laboratory. Subsequent reactions have been well characterized. After a downward shift in temperature or a decrease in membrane fluidity, the expression of the genes for desaturases is enhanced and the desaturases are synthesized *de novo* and targeted to both the plasma membrane and the thylakoid membranes (Mustardy *et al.*, 1996). These enzymes catalyze the desaturation of the fatty acids in the membrane lipids to compensate for the decrease in membrane fluidity that has been caused by cold stress.

Two Histidine Kinases Involved in the Perception and Transduction of Cold Signal

Histidine kinases are involved in the perception of chemical and osmotic stress in prokaryotes, yeast and plants. Therefore, they might also play an important role in the perception and transduction of cold signals. The genome of *Synechocystis* sp. PCC 6803 (Kaneko *et al.*, 1995; 1996) includes 43 putative genes for histidine kinases (Mizuno *et al.*, 1996), all of which are potential candidates for components of perception and transduction pathways for environmental signals.

In order to identify sensors and tranducers of cold signals, we generated a strain of *Synechocystis* sp. PCC 6803, in which the promoter region of the cold-inducible *desB* gene was ligated with the coding region of the *lux* gene for a bacterial luciferase (Los *et al.*, 1997). We inactivated separately each of the genes for histidine kinases by inserting an antibiotic-resistance cartridge, creating a "gene-knockout" library. We screened members of the

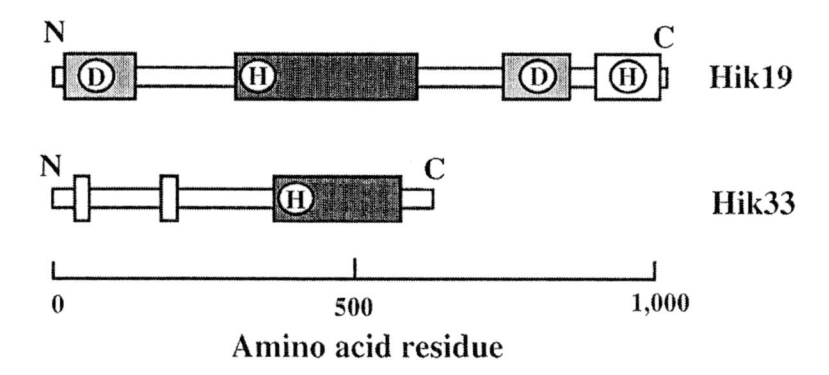

Figure 7. Schematic Representation of Structures of Hik19 and Hik33. H and D indicate histidine and aspartic acid residues, respectively. In Hik19, the hatched boxes, dark box, and open box indicate signal-receiver domains, a histidine kinase domain, and a histidine phospho-transfer domain, respectively. In Hik33, the open boxes and dark box indicate membrane-spanning and histidine kinase domains, respectively.

library for loss of cold inducibility by monitoring luciferase activity at a low temperature. The results suggested that two histidine kinase, Hik19 and Hik33, might be involved in the perception and transduction of cold signals (Suzuki et al., 2000).

Figure 7 shows a schematic representation of the structures of Hik19 and Hik33. Hik33 contains a histidine kinase domain and two membrane-spanning domains. Since the cold sensor might be located in the plasma membrane of cyanobacterial cells, it seems likely that the activity of Hik33 might be sensitive to changes in membrane fluidity, and, thus, might be a sensor of temperature. Hik19 contains a histidine kinase domain, two signal-receiver domains, and a histidine phospho-transfer domain (Hpt-domain), but no membrane-associated domain. Thus, Hik19 is more likely than Hik33 to be a transducer of cold signals.

Acknowledgement

This work was supported, in part, by a Grant-in-Aid for Specially Promoted Research (no. 081202011) from the Ministry of Education, Science and Culture, Japan, to N.M.

References

Campbell, D., Zhou, G., Gustafsson, P., Öquist, G., and Clarke, A. K. 1995. Electron transport regulates exchange of two forms of photosystem II D1 protein in the cyanobacterium *Synchococcus*. EMBO J. 14: 5457-5466.

Celerin, M., Gilpin, A.A., Schisler, N.J., Ivanov, A.G., Miskiewicz, E., Krol, M., and Laudenbach, D.E. 1998. ClpB in a cyanobacterium: predicted structure, phylogenetic relationships, and regulation by light and temperature. J. Bacteriol. 180: 5173-5182.

Chamot, D., Magee, W.C., Yu, E., and Owttrim, G.W. 1999. A cold shock-induced cyanobacterial RNA helicase. J. Bacteriol. 181: 1728-1732.

Cossins, A.R. 1994. Homeoviscous adaptation of biological membranes and its functional significance. In: Temperature Adaptation of Biological Membranes. A.R. Cossins, ed. Portland Press, London. p. 63-76.

Cohen, M.F., Wallis, J.G., Campbell, E.L., and Meeks, J.C. 1994. Transposon mutagenesis of *Nostoc* sp. strain ATCC 29133, a filamentous cyanobacterium with multiple cellular differentiation alternatives. Microbiol. 140: 3233-3240.

Dolganov, N., and Grossman, A.R. 1993. Insertional inactivation of genes to isolate mutants of *Synechococcus* sp. strain PCC 7942: isolation of filamentous strains. J. Bacteriol. 175: 7644-7651.

Elhai, J., and Wolk, C.P. 1988. Conjugal transfer of DNA to cyanobacteria. Methods Enzymol. 167: 747-765.

Eriksson, M.-J., and Clarke, A.K. 1996. The heat shock protein Cl pB mediates the development of thermotolerance in the cyanobacterium *Synechococcus* sp. strain PCC 7942. J. Bacteriol. 178: 4839-4846.

Fraser, C.M., Gocayne, J.D., White, O., Adams, M.D., Clayton, R.A., Fleischmann, R.D., Bult, C.J., Kerlavage, A.R., Sutton, G., Kelley, J.M., Fritchman, J.L., Weidman, J.F., Small, K.V., Sandusky, M., Fuhrmann, J., Nguyen, D., Utterback, T.R., Saudek, D.M., Phillips, C.A., Merrick, J.M., Tomb, J.-F., Dougherty, B.A., Bott, K.F., Hu, P.-C., Lucier, T.S., Peterson, S.N., Smith, H.O., Hutchison, III, C.A., and Venter, J.C. 1995. The minimal gene complement of *Mycoplasma genitalium*. Science. 270: 397-403.

Fuller-Pace, F.V. 1994. RNA helicases: modulators of RNA structure. Trends Cell Biol. 4: 271-274.

Glatz, A., Vass, I., Los, D.A., and Vígh, L. 1999. The *Synechocystis* model of stress: From molecular chaperones to membranes. Plant. Physiol. Biochem. 37: 1-12.

Gombos, Z., Wada, H., and Murata, N. 1992. Unsaturation of fatty acids in membrane lipids enhances tolerance of the cyanobacterium *Synechosystis* PCC6803 to low-temperature photoinhibition. Proc. Natl. Acad. Sci. USA. 89: 9959-9963.

Gombos, Z., Wada, H., and Murata, N. 1994. The recovery of photosynthesis from low-temperature photoinhibition is accelerated by the unsaturation of membrane lipids: a mechanism of chilling tolerance. Proc. Natl. Acad. Sci. USA. 91: 8787-8791.

Gombos, Z., Kanervo, E., Tsvetkova, N., Sakamoto, T., Aro, E.-M., and Murata, N. 1997. Genetic enhancement of the ability to tolerate photoinhibition by introduction of unsaturated bonds into membrane glycerolipids. Plant Physiol. 115: 551-559.

Gorbalenya, A.E., and Koonin, E.V. 1993. Helicases: amino acid sequence comparisons and structure-function relationships. Curr. Opin. Struct. Biol. 3: 419-429.

Gray, J.E., and Herson, D.S. 1976. Functional 70S hybrid ribosomes from blue-green algae and bacteria. Arch. Microbiol. 109: 95-99.

Harris, E.H., Boynton, J.E., and Gillham, N.W. 1994. Chloroplast ribosomes and protein synthesis. Microbiol. Rev. 58: 700-754.

Haselkorn, R. 1991. Genetic systems in cyanobacteria. Methods Enzymol.

204: 418-430.

Held, W.A., Ballou, B., Mizushima, S., and Nomura, M. 1974. Assembly mapping of 30S ribosomal proteins from *Escherichia coli*. J. Biol. Chem. 249: 3103-3111.

Higashi, S., and Murata, N. 1993. An *in vivo* study of substrate specificities of acyl-lipid desaturases and acyltransferases in lipid synthesis in *Synechocystis* PCC6803. Plant Physiol. 102: 1275-1278.

Ishizaki-Nishizawa, O., Fujii, T., Azuma, M., Sekiguchi, K., Murata, N., Ohtani, T., and Toguri, T. 1996. Low-temperature resistance of higher plants is significantly enhanced by a nonspecific cyanobacterial desaturase. Nature Biotech. 14: 1003-1006.

Jones, P.G., and Inoue, M. 1994. The cold-shock response − a hot topic. Mol. Microbiol. 11: 811-818.

Jones, P.G., Mitta, M., Kim, Y., Jiang, W., and Inoue, M. 1996. Cold shock induces a major ribosomal-associated protein that unwinds double-stranded RNA in *Escherichia coli*. Proc. Natl. Acad. Sci. USA. 93: 76-80.

Kaneko, T., and Tabata, S. 1997. Complete genome structure of the unicellular cyanobacterium *Synechocystis* sp. PCC6803. Plant Cell Physiol. 38: 1171-1176.

Kaneko T., Sato S., Kotani H., Tanaka A., Asamizu E., Nakamura Y., Miyajima N., Hirosawa M., Sugiura M., Sasamoto S., Kimura T., Hosouchi T., Matsuno A., Muraki A., Nakazaki N., Naruo K., Okumura S., Shimpo S., Takeuchi C., Wada T., Watanabe A., Yamada M., Yasuda M., and Tabata S. 1996. Sequence analysis of the genome of the unicellular cyanobacterium *Synechocystis* sp. strain PCC 6803, II. Sequence determination of the entire genome and assignment of potential protein-coding regions. DNA Res. 3: 109-136.

Kanervo, E., Tasaka, Y., Murata, N., and Aro, E.-M. 1997. Membrane lipid unsaturation modulated processing of photosystem II reaction-center protein D1 at low temperature. Plant Physiol. 114: 841-849.

Kanervo, E., Murata, N., and Aro, E.-M. 1998. Massive breakdown of the photosystem II polypeptides in a mutant of the cyanobacterium *Synechocystis* sp. PCC 6803. Photosynth. Res. 57: 81-91.

Kates, M., Pugh, E.L., and Ferrante, G. 1984. Regulation of membrane fluidity by lipid desaturases. Biomembranes. 12: 379-395.

Kenan, D.J., Query, C.C., and Keene, J.D. 1991. RNA recognition: towards identifying determinants of specificity. Trends Biochem. Sci. 16: 214-220.

Kessel, M., Maurizi, M.R., Kim, B., Kocsis, E., Trus, B.L., Singh, S.K., and Steven, A.C. 1995. Homology in structural organization between *E. coli* ClpAP protease and the eukaryotic 26S proteasome. J. Mol. Biol. 250: 587-594.

Kunst, F., Ogasawara, N., Moszer, I., Albertini, A.M., Alloni, G., Azevedo, V., Bertero, M.G., Bessières, P., Bolotin, A., Borchert, S., Borriss, R., Boursier, L., Brans, A., Braun, M., Brignell, S.C., Bron, S., Brouillet, S., Bruschi, C.V., Caldwell, B., Capuano, V., Carter, N.M., Choi, S.-K., Codani, J.-J., Connerton, I.F., Danchin, A., et al. 1997. The complete genome sequence of Gram-positive bacterium *Bacillus subtilis*. Nature. 390: 249-256.

Labarre, J., Chauvat, F., and Thuriaux, P. 1989. Insertional mutagenesis by random cloning of antibiotic resistance genes into the genome of the cyanobacterium *Synechocystis* strain PCC 6803. J. Bacteriol. 171: 3449-3457.

Los, D.A., and Murata, N. 1994. The low-temperature-induced accumulation of the *desA* transcript in *Synechocystis* PCC6803 is a result of both activation of transcription and maintenance of RNA stability. Russian J. Plant Physiol. 41: 146-151.

Los, D.A., and Murata, N. 1998. Structure and expression of fatty acid desaturases. Biochim. Biophys. Acta. 1394: 3-15.

Los, D.A., Horvàth, I., Vigh, L., and Murata, N. 1993. The temperature-dependent expression of the desaturase gene *desA* in *Synechocystis* PCC6803. FEBS Lett. 318: 57-60.

Los, D.A., Ray, M. K., and Murata, N. 1997. Differences in the control of the temperature-dependent expression of four genes for desaturases in *Synechocystis* sp. PCC 6803. Mol. Microbiol. 25: 1167-1176.

Margulis, L. 1975. Symbiotic theory of the origin of eukaryotic organelles; criteria for proof. Symp. Soc. Exp. Biol. 29: 21-38.

Maurizi, M.R., Clark, W.P., Katayama, Y., Rudikoff, S., Pumphrey, J., Bowers, B., and Gottesman, S. 1990a. Sequence and structure of Clp P, the proteolytic component of the ATP-dependent Clp protease of *Escherichia coli*. J. Biol. Chem. 265: 12536-12545.

Maurizi, M.R., Clark, W.P., Kim, S.-H., and Gottesman, S. 1990b. Clp P represents a unique family of serine proteases. J. Biol. Chem. 265: 12546-12552.

Meeks, J.C., and Castenholz, R.W. 1971. Growth and photosynthesis in an extreme thermophile, *Synechococcus lividus* (Cyanophyta). Arch. Mikrobiol. 78: 25-41.

Mizuno, T., Kaneko, T., and Tabata, S. 1996. Compilation of all genes encoding bacterial two-component signal transducers in the genome of the cyanobacterium, *Synechocystis* sp. strain PCC 6803. DNA Res. 3: 407-414.

Murata, N. 1989. Low-temperature effects on cyanobacterial membranes. J. Bioenerg. Biomembr. 21: 61-75.

Murata, N., and Los, D.A. 1997. Membrane fluidity and temperature perception. Plant Physiol. 115: 875-879.

Murata, N., and Wada, H. 1995. Acyl-lipid desaturases and their importance in the tolerance and acclimatization to cold of cyanobacteria. Biochem. J. 308: 1-8.

Murata, N., Wada, H., and Gombos, Z. 1992. Modes of fatty-acid desaturation in cyanobacteria. Plant Cell Physiol. 33: 933-941.

Mustardy, L., Los, D.A., Gombos, Z., and Murata, N. 1996. Immunocytochemical localization of acyl-lipid desaturases in cyanobacterial cells: Evidence that both thylakoid membranes and cytoplasmic membranes are sites of lipid desaturation. Proc. Natl. Acad. Sci. USA. 93: 10524-10527.

Nagai, K., Oubridge, C., Ito, N., Avis, J., and Evans, P. 1995. The RNP domain: a sequence-specific RNA-binding domain involved in processing and transport of RNA. Trends Biochem. Sci. 20: 235-240.

Nakamura, Y., Kaneko, T., Hirosawa, M., Miyajima, N., and Tabata, S. 1998. CyanoBase, a www database containing the complete nucleotide sequence of the genome of *Synechocystis* sp. strain PCC6803. Nucleic Acids Res. 26: 63-67.

Nicol, S.M., and Fuller-Pace, F.V. 1995. The "DEAD box" protein DbpA interacts specifically with the peptidyltransferase center in 23S rRNA. Proc. Natl. Acad. Sci. USA. 92: 11681-11685.

Nishi, K., Morel-Deville, F., Hershey, J.W.B., Leighton, T., and Schnier, J.A. 1988. An eIF-4A-like protein is a suppressor of an *Escherichia coli* mutant defective in 50S ribosomal subunit assembly. Nature. 336: 496-498.

Porankiewicz, J., and Clarke, A.K. 1997. Induction of a heat shock protein, ClpB, affects cold acclimation in the cyanobacterium *Synechococcus* sp. strain PCC 7942. J. Bacteriol. 179: 5111-5117.

Porankiewicz, J., Schelin, J., and Clarke, A.K. 1998. The ATP-dependent Clp protease is essential for acclimation to UV-B and low temperature in the cyanobacterium *Synechococcus*. Mol. Microbiol. 29: 275-283.

Priscu, J.C., Fritsen, C.H., Adams, E.E., Giovannoni, S.J., Paerl, H.W., McKay, C.P. Doran, P.T., Gordon, D.A., Lanoil, B.D., and Pinckney, J.L. 1998. Perennial Antarctic lake ice: an oasis for life in a polar desert. Science. 280: 2095-2097.

Psenner, R., and Sattler, B. 1998. Life at the freezing point. Science. 280: 2073-2074.

Py, B., Higgins, C.F., Krisch, H.M., and Carpousis, A.J. 1996. A DEAD-box RNA helicase in the *Escherichia coli* RNA degradosome. Nature. 381: 169-172.

Sakamoto, T., and Bryant, D.A. 1997. Temperature-regulated mRNA accumulation and stabilization for fatty acid desaturase genes in the cyanobacterium *Synechococcus* sp. strain PCC 7002. Mol. Microbiol. 23: 1281-1292.

Sakamoto, T., Higashi, S., Wada, H., Murata, N., and Bryant, D.A. 1997. Low-temperature-induced desaturation of fatty acids and expression of desaturase genes in the cyanobacterium *Synechococcus* sp. PCC 7002. FEMS Microbiol. Lett. 152: 313-320.

Sato, N. 1992. Cloning of a low-temperature-inducible gene *lti2* from the cyanobacterium *Anabaena variabilis* M3 that is homologous to α-amylases. Plant Mol. Biol. 18: 165-170.

Sato, N. 1994. A cold-regulated cyanobacterial gene cluster encodes RNA-binding protein and ribosomal protein S21. Plant Mol. Biol. 24: 819-823.

Sato, N. 1995. A family of cold-regulated RNA-binding protein genes in the cyanobacterium *Anabaena variabilis* M3. Nucleic Acids Res. 23: 2161-2167.

Sato, N., and Murata, N. 1980. Temperature shift-induced responses in lipids in the blue-green alga, *Anabaena variabilis*: the central role of diacylmonogalactosylglycerol in thermo-adaptation. Biochim. Biophys. Acta. 619: 353-366.

Sato, N., and Maruyama, K. 1997. Differential regulation by low temperature of the gene for an RNA-binding protein, *rbpA3*, in the cyanobacterium

Anabaena variabilis strain M3. Plant Cell Physiol. 38: 81-86.

Sato, N., and Nakamura, A. 1998. Involvement of the 5'-untranslated region in cold-regulated expression of the *rbpA1* gene in the cyanobacterium *Anabaena variabilis* M3. Nucleic Acids Res. 26: 2192-2199.

Sato, N., Murata, N., Miura, Y., and Ueta, N. 1979. Effect of growth temperature on lipid and fatty acid compositions in the blue-green algae *Anabaena variabilis* and *Anacystis nidulans*. Biochim. Biophys. Acta. 572: 19-28.

Sato, N., Tachikawa, T., Wada, A., and Tanaka, A. 1997. Temperature-dependent regulation of the ribosomal small-subunit protein S21 in the cyanobacterium *Anabaena variabilis* M3. J. Bacteriol. 179: 7063-7071.

Sato, N., Wada, A., and Tanaka, A. 1998. Ribosomal proteins in the cyanobacterium *Anabaena variabilis* strain M3: presence of L25 protein. Plant Cell Physiol. 39: 1367-1371.

Schnuchel, A., Wiltscheck, R., Czisch, M., Herrler, M., Willimsky, G., Graumann, P., Marahiel, M.A., and Holak, T.A. 1993. Structure in solution of the major cold-shock protein from *Bacillus subtilis*. Nature. 364: 169-171.

Schopf, J.W., Barghoorn, E.S., Maser, M.D., and Gordon, R.O. 1965. Electron microscopy of fossil bacteria two billion years old. Science. 149: 1365-1367.

Schwartz, S.H., Black, T.A., Jäger, K., Panoff, J.-M., and Wolk, C.P. 1998. Regulation of an osmoticum-responsive gene in *Anabaena* sp. strain PCC 7120. J. Bacteriol. 180: 6332-6337.

Sippola, K., Kanervo, E., Murata, N., and Aro, E.-M. 1998. A genetically engineered increase in fatty acid unsaturation in *Synechococcus* sp. PCC 7942 allows exchange of D1 protein forms and sustenance of photosystem II activity at low temperature. Eur. J. Biochem. 251: 641-648.

Suzuki, I., Los, D. A., Kanesaki, Y., Mikami, K., and Murata, N. 2000. The pathway for perception and transduction of low-temperature signals in *Synechocystis*. EMBO J. 19: 1327-1334.

Tasaka, Y., Gombos, Z., Nishiyama, Y., Mohanty, P., Ohba, T., Ohki, K., and Murata, N. 1996. Targeted mutagenesis of acyl-lipid desaturases in *Synechocystis*: Evidence for the important roles of polyunsaturated membrane lipids in growth, respiration and photosynthesis. EMBO J. 15: 6416-6425.

Thieringer, H.A., Jones, P.G., and Inoue, M. 1998. Cold shock and adaptation. Bioessays. 20: 49-57.

Thompson, M.W., and Maurizi, M.R. 1994. Activity and specificity of *Escherichia coli* ClpAP protease in cleaving model peptide substrates. J. Biol. Chem. 269: 18201-18208.

Vermaas, W.F. 1998. Gene modifications and mutation mapping to study the function of photosystem II. Methods Enzymol. 297: 293-310.

Vigh, L., Los, D.A., Horvath, I., and Murata, N. 1993. The primary signal in the biological perception of temperature: Pd-catalyzed hydrogenation of membrane lipids stimulated the expression of the *desA* gene in *Synechocystis* PCC6803. Proc. Natl. Acad. Sci. USA. 90: 9090-9094.

Wada, H., Gombos, Z., and Murata, N. 1990. Enhancement of chilling tolerance of a cyanobacterium by genetic manipulation of fatty acid desaturation. Nature. 347: 200-203.

Wada, H., Gombos, Z., and Murata, N. 1994. Contribution of membrane lipids to the ability of the photosynthetic machinery to tolerate temperature stress. Proc. Natl. Acad. Sci. USA. 91: 4273-4277.

Ward, D.M., Ferris, M.J., Nold, S.C., and Bateson, M.M. 1998. A natural view of microbial biodiversity within hot spring cyanobacterial mat communities. Microbiol. Mol. Biol. Rev. 62: 1353-1370.

Williams, J.G.K. 1988. Construction of specific mutations in photosystem II photosynthetic reaction center by genetic engineering methods in *Synechocystis* PCC6803. Methods Enzymol. 167: 766-778.

Williams J.G., and Szalay, A.A. 1983. Stable integration of foreign DNA into the chromosome of the cyanobacterium *Synechococcus* R2. Gene. 24:37-51.

Ye, L., and Sugiura, M. 1992. Domains required for nucleic acid binding activities in chloroplast ribonucleoproteins. Nucleic Acids Res. 20: 6275-6279.

6

Molecular Responses of Plants to Cold Shock and Cold Acclimation

Charles Guy

Plant Molecular and Cellular Biology Program,
Department of Environmental Horticulture,
University of Florida, Gainesville, Florida 32611-0670, USA

Abstract

The Plant Kingdom encompasses a grouping of mostly sessile organisms that show extreme variation in morphology, size, ecological adaptation, life cycle, and climatic tolerance. With the exception of low elevation tropical environments, plants living just about anywhere else in the world may be subject to temperatures below that which are optimal for growth and survival. Consequently, the range of tolerance to low temperature stress in the Plant Kingdom is as great as the natural variation in low temperatures. For mesophilic plants, sub-optimal low temperature could range from 15°C down to -55°C. In the past 10 years, more than 100 genes have been shown to be preferentially expressed in response to low temperatures. Significant progress in understanding the responses of plants to low temperature has occurred in the areas of signal perception and transduction pathways, transcriptional control and the characterization of a variety of stress-related proteins. A common aim of much of the research on cold stress in plants is to find ways to enhance the stress tolerance and reduce economic losses.

Introduction

Imagine for a moment that your feet are anchored to the ground and you are standing in St. Paul, Alto Rio Senguerr, Torino or Sapporo and it is summer. You are outside and can't go inside. Now imagine having to remain in that place for the entire year; all of your life. This is the life of a tree. Now these four cities are approximately half way from the equator to either the North or South Pole. Some have more extreme temperatures than others, primarily in the winter. During the summer the warmest air temperature might be about

40°C and in winter the coldest might be as low as -35°C. That is a mere 75°C range that the above ground portions of trees in these locales must be able to withstand in order to continue to live.

The upper 30°C of this temperature range present few problems for the tree, but the lower 45°C of the range can be quite problematic. From about 10°C down to 0°C represents a range of temperatures where metabolic activity is strongly reduced largely from the influence of temperature on the kinetic parameters of enzyme catalyzed reactions. For trees that have become adapted to the climatic conditions at these selected sites, specific modifications to the enzymatic complement of cells overcomes the metabolic challenge of sub-optimal temperatures. What are some of the modifications that occur? In some cases, the cell produces more enzyme (Guy et al., 1992a) to maintain adequate activity, increases the activation state (Holaday et al., 1992; Hurry et. al., 1994), and in other cases, new isozymic forms with enhanced catalytic function at low temperature are made (Guy and Carter, 1984). However, the latter response does not always hold true as in the case of phenylalanine ammonia lyase isozymes of parsley where all members of the family have equivalent kinetic characteristics (Appert et al., 1994). Perhaps during the long period of adaptation to these particular locales and climates, the stability of the folded state of certain proteins and their biogenesis pathways has become optimized for the sub-optimal temperature conditions (Guy et al., 1997). Changes necessary to maintain cellular homeostasis are not limited solely to changes in the enzymatic complement, but may also include a number of macromolecular processes from protein synthesis (Guy et al., 1985), to membrane structural (Uemura et al., 1995) and functional (Uemura and Steponkus, 1994) stability to the structural organization and stability of chromatin (Koukalova et al., 1997; Mineur et al., 1998).

Every organism has a range of temperatures that is optimal for growth processes. Above or below this range growth begins to diminish and at some point, depending on the organism, or more specifically the plant in question, the divergence from the optimum range begins to have more deleterious consequences. This dynamic response to temperature can be generalized to illustrate its profound influence on plants (Figure 1). For plants, photosynthesis has been for obvious reasons, inextricably associated with overall health and vigor (Strand et al., 1999). Photosynthesis, but more importantly net carbon assimilation, as well as respiratory activities, are temperature dependent processes. As temperature declines or increases from an organism dependent optimum, determined as a result of adaptation to a particular environment over evolutionary time, carbon assimilation and respiration will similarly decline or increase. Eventually, the deviation from the optimum temperature will drive the rates of these basic processes to zero. However, over the long term, the gross photosynthesis and respiratory processes maybe less important than the net photosynthesis which is the difference between overall carbon assimilation and respiratory consumption. Under conditions that lead to zero net photosynthesis, starvation is guaranteed. However, this level of deviation from the optimum temperature

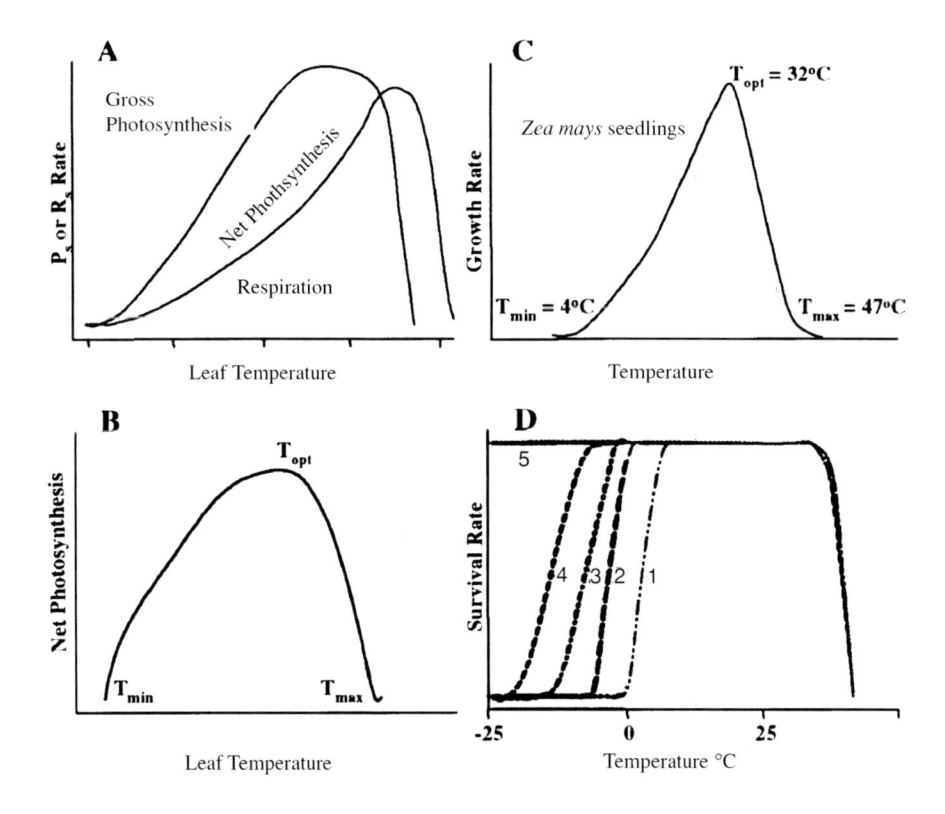

Figure 1. Temperature Influence. Influence of temperature on photosynthesis, respiration, growth and survival of plants. Numbers denote different low temperature survival profiles. Adapted from Pisek et al., 1973 and Lundegårdh, 1954).

may not yet be sufficient to result in an immediate direct lethal injury during a short-term stress, but may require a more extreme temperature or longer duration of stress.

Inducible Low Temperature Stress Tolerance

The diversity in tolerance to low temperature extremes represented by different acclimation states and/or different mesophilic plants are conceptualized in panel D of Figure 1. When plants are exposed to a nonlethal high temperature for a short duration (from minutes to a few hours), survival upon exposure to an otherwise lethal higher temperature is increased (Altschuler and Mascarenhas, 1982). Usually the increase in thermotolerance resulting from acquired thermotolerance, while significant, is modest and ranges from 1-4°C. In contrast, mesophilic plants as a group can exhibit extreme diversity in low temperature tolerance represented by at least five different survival profiles. Plants that show injury and/or loss of viability at temperatures between 0°C and 12°C are considered to be chilling sensitive (line 1 in Figure 1D). A few examples would include rice (Tajima et al., 1983), corn (Taylor et al., 1974), african violet and *Coffea* (Bodner and Larcher, 1987). The next most cold sensitive group is not injured by exposure to low nonfreezing temperatures, but is immediately injured or killed when ice begins to form inside their tissues (line 2). Perhaps the best example would be the cultivated potato *Solanum tuberosum* (Sukumaran, and Weiser, 1972) which tolerates nonfreezing temperatures well, but is damaged by the slightest frost. The next level of hardiness would be plants that can withstand ice formation in their tissues, but are killed at high sub-zero temperatures (-6°C to -1°C) (line 3). Examples of this grouping would be petunia and most members of the genus *Citrus*, which are killed at temperatures of -3°C to -6°C (Yelenosky and Guy, 1989). The next hardiness level would be plants that can, when acclimated, survive freezing from temperatures ranging from about -10°C to -30°C (line 4). Many of the cereals, temperate herbaceous species and fruit producing trees fall into this category (Fowler and Gusta, 1979; Scorza et al., 1983). The most hardy plants, and the last grouping, can survive -30°C to -50°C in nature, and immersion in liquid nitrogen (-196°C) in the lab when fully acclimated (line 5). Numerous temperate and alpine trees fit into this group and a famous example of this is black locust (*Robinia pseudoacacia*) (Sakai and Yoshida, 1968). In the nonacclimated state (in spring after growth has resumed and until growth ceases in late summer), most of the plants that fall into the two most hardy categories (lines 4, 5) exhibit a marked sensitivity to freezing stress that in most instances would be characterized by the profiles represented by lines 2 or 3 (Figure 1D).

The plants described above would be considered mesophilic, at least with respect to the temperature range (15°C to 35°C) for optimal growth. For the vast majority of such plants from the small annual herbaceous species to the largest perennial trees, the upper limits of survival to short term high temperatures range from roughly 40°C - 55°C. This upper limit, however, is strongly influenced by at least three factors; duration of the stress, environmental conditions prior to the heat stress, and developmental stage (Khan, 1976). These same factors are also important in low temperature stress survival.

Low Temperature Stress Proteins

The response of plants to low temperature in many ways is analogous to the response to heat shock, but not identical. When a plant is subjected to a nonlethal heat stress for a short duration, the ability of the plant to survive higher temperatures is the result of an inducible acquired thermotolerance (Neumann et al., 1989). There is a large body of correlative evidence to suggest that the synthesis of heat shock proteins is necessary for the increased survival that results from acquired thermotolerance (Neumann et al., 1989; Vierling, 1991), although there is other evidence that heat shock proteins alone are not sufficient (Singer and Lindquist, 1998). Presently 11 different classes of heat shock proteins are recognized in plants (Nover and

Scharf, 1997). In general terms, the major functions of most heat shock proteins are now known, and 10 out of the 11 classes are involved in protein metabolism either during non-stressful conditions and/or during heat stress. Consequently, they can be classified as molecular chaperones (Nover and Scharf, 1997).

Exposure to a low non-lethal temperature usually results in an acclimation response that is characterized by a greater ability to resist injury or survive a low temperature stress that otherwise would be lethal (Levitt, 1972). This process known as cold acclimation, illustrated in Figure 1, is similar to the acquired thermotolerance of the heat shock response. Over the last 10 years, as with the heat shock response, changes in gene expression and the synthesis of cold shock or cold acclimation proteins has been frequently correlated with enhanced cold tolerance (Hughes and Dunn, 1996; Thomashow, 1998; Thomashow, 1999).

The pattern of protein synthesis during cold acclimation is very dissimilar to the heat shock response. Most of the housekeeping proteins synthesized in the absence of stress continue to be produced (Guy and Haskell, 1987), along with the synthesis of several cold stress proteins (actually several hundred may be induced). In contrast to heat shock, the vast majority of the cold stress proteins do not belong to the classical molecular chaperones grouping (Hughes and Dunn, 1996; Thomashow, 1998), nor is the response as conserved as one might suspect given the differences inherent in the tolerance capacities of plants illustrated by Figure 1. Nevertheless, studies with plants have identified several cold stress proteins belonging to the various classes of heat shock proteins, such as Hsps104 and 90 (Pareek *et al.,* 1995), two members of the Hsp70 family (Wang and Fang, 1996; Muench *et al.,* 1997), and two members of the smHsp group (Yeh *et al.,* 1995; Lee *et al.,* 1995). In contrast, other cold stress responsive genes or proteins represent just about any type and function imaginable (Hughes and Dunn, 1996; Thomashow, 1998) such as alternative oxidases (Ito *et al.,* 1997), a vacuolar proton translocating pyrophosphatase (Carystinos *et al.,* 1995), an omega-3 fatty acid desaturase (Kodama *et al.,* 1997), a water stress/late embryogenesis abundant protein (Takahashi *et al.,* 1994) and a cell cycle related gene (Kidou *et al.,* 1994) to highlight a few. However, certain major classes of cold regulated genes are of particular interest at this time and are described in greater detail.

Antifreeze Proteins

Antifreeze proteins (AFPs) were first discovered and characterized in polar fishes (DeVries, 1971) that inhabit waters where the temperature is frequently in the -1°C to -2°C range. AFPs are present in relatively high concentration and have the ability to inhibit the growth of an ice crystal by preventing the accretion of water molecules to the growing faces of the crystal by a noncolligative mechanism. AFPs exhibit a property of thermal hysteresis where the temperature that promotes freezing (ice crystal growth) is lower than the temperature that promotes thawing. Consequently, antifreeze proteins are frequently synonymous with thermal hysteresis proteins (THPs). Most of the fish antifreezes show modest hysteresis (Burcham et al., 1986). While the polar marine environment poses a problem for organisms to keep low osmolal bodily fluids from forming ice nuclei and growing damaging ice crystals, the differential between the freezing point depression of bodily fluids and the surrounding temperatures is relatively minimal.

Given the vastly more variable environments experienced by terrestrial plants, many plant cryobiologists considered that plants would not derive significant benefit from AFPs that could only arrest ice formation over a very narrow temperature range. This however, changed with the discovery of THPs in plants (Griffith et al., 1992; Urrutia, et al., 1992). Protein mixtures found in the apoplast (extracellular regions of the tissue) of rye plant leaves were shown to modify the growth pattern of ice crystals and depressed the freezing temperature of aqueous solutions by a noncolligative mechanism (Griffith et al., 1992). This was the first unambiguous demonstration that proteins with antifreeze-like properties were present in plants. At almost the same time, the presence of proteins in extracts from plant tissues with characteristics of conferring thermal hysteresis on aqueous solutions was reported (Urrutia, et al., 1992). Additional studies provided evidence for THPs in several different species of plants bringing the total with thermal hysteresis activity to 23 representing diverse phylogenetic groups (Duman and Olsen, 1993). That AFPs were the basis for the hysteric activity was confirmed when the rye apoplast proteins were purified and shown to have antifreeze activity. A total of seven proteins exhibited antifreeze activity. None of the proteins were cross-reactive with antisera to fish antifreeze proteins suggesting they may represent a new or novel class of AFPs (Hon et al., 1994).

The most abundant proteins present in the apoplast of acclimated rye leaves were determined to have antifreeze activity. Using peptide sequencing, and specific antisera, the antifreeze proteins were found to belong to three classes of pathogenesis-related (PR) proteins, endochitinases (CLP), endo-ß-1,3-glucanases (GLP) and thaumatin-like proteins (TLP) (Hon et al., 1995). In contrast, the freeze-sensitive tobacco did not produce apoplastic pathogenesis-related proteins of the same classes with antifreeze activity after low temperature exposure indicating that AFPs were not common to all plants. In rye leaves, the proteins were strongly accumulated when the plants were given a cold acclimation treatment at 5°C/2°C for seven weeks. Additional surveys showed that AFP activity was present only in the apoplastic extracts of freezing tolerant monocotyledons after cold acclimation (Antikainen and Griffith, 1997). Their accumulation in wheat at low temperature appears to be regulated by chromosome 5 (Griffith et al., 1997). More recently, Chun et al., (1998) using wheat chromosome substitution lines and the parental lines Chinese Spring and Cheyenne, demonstrated that Cheyenne chromosomes 5B and 5D have major influence on the accumulation of antifreeze activity and AFPs.

In an interesting twist, a protein from carrot was found to have characteristics associated with antifreeze proteins (Worrall et al., 1998; Meyer *et al.,* 1999). When the carrot protein was expressed in transgenic tobacco or Arabidopsis plants, the protein retained antifreeze properties in that it inhibited the recrystallization process or altered the morphology of ice crystal growth. The protein conferred a thermal hysteresis on extracts that was seemingly unaltered by degylcosylation. Instead of being a member of the three PR families as proteins of the rye AFPs, the carrot protein had similarity to polygalacturonase inhibitor proteins (PGIPs) whose purpose is considered to inhibit fungal polygalacturonases. It has a leucine-rich motif. Such proteins contain repeated units of a 24-residue motif that has regularly spaced leucines. Along with the rye AFPs, the carrot protein appears to be a second example of where a protein with one function evolves a new and unrelated function. While the AFP function may be evolutionarily recent, it is reasonable to consider that these AFP/PR proteins actually have a dual function to protect the plants against ice recrystallization and pathogen attack during the unfavorable conditions of winter.

Interestingly, disease resistance and freezing tolerance seem to be related to the accumulation of the PR-like proteins. Cold acclimation of many freeze-tolerant grasses enhances resistance to snow molds, powdery mildews, leaf spots and rusts (Hon *et al.,* 1995). Immunolocalization of GLP, CLP and TLP showed the proteins are present in the epidermis and cells adjacent to intercellular spaces (Antikainen *et al.,* 1996). In addition, the AFPs accumulated in cold acclimated plants in mesophyll cell walls, secondary cell walls of xylem vessels and epidermal cell walls (Griffith *et al.,* 1997). Apparently different forms of these same PR-proteins were present in nonacclimated plants although they lacked antifreeze properties, and they were most abundant in different locations. It appears that low temperature specific forms are localized and accumulated in response to low temperature in ways that reflect a potential common pathway for ice and pathogens to enter the tissue. Apparently accumulation of AFPs during cold acclimation is a relatively specific response and not general for all plants. In an unexpected reversal, Xu and colleagues (Xu *et al.,* 1998) found the rhizobacterium *Pseudomonas putida* GR12-2 produces an antifreeze protein that is secreted into the growth medium. When purified, the protein was found to have properties similar to the bacterial ice nucleation protein (Wolber *et al.,* 1986). It appears that this protein also has low amounts of ice-nucleation activity, which seems paradoxical for an AFP.

Cold Shock Domain Proteins

A superfamily of proteins in eukaryotic organisms contain the cold-shock domain (Graumann and Maraheil 1998), and is found in proteins that bind nucleic acids. They were first described as Y-box binding proteins that bind to the Y-box, a *cis*-element with the sequence CCAAT (Didier *et al.,* 1988). The cold shock domain name comes from the similarity of Y-box type proteins with the sequences found in CSPA (Goldstein *et al.,* 1990) and other known bacterial CSPs. In addition, the cold shock domain is known to mediate RNA-binding in animals and bacteria (Murray, 1994; Jiang *et al.,* 1997), and

```
E.c. CSPA:     7 GIVKWFNADKGFGFITPDDGSKDVFVHFSAIQND---GYKSLDEGQKVSFTIESGAKG   61 (M30139)
N.s. GRP2:    11 GtVKWFsdqKGFGFITPDDGgeDlFVHqSgIrse---GfrSLaEGetVeFevESGgdG   65 (X60007)
A.t. GRP2:    13 GsVKWFdtqKGFGFITPDDGgdDlFVHqSsIrse---GfrSLaaeeaVeFevE         62 (S47408)
A.t. GRP2b:   17 GtVKWFdtqKGFGFITPsDGgdDlFVHqSsIrse---GfrSLaaeesVeFdvEvdnsG   71 (U39072)
A.t. GRPzf:   11 GkVsWFsdgKGyFGFITPDDGgeelFVHqSsIvsD---GfrSLtlGesVeyeIalGsdG  65 (AC003952)
D.m. Y-box:   65 GtVKWFNvksGyGFInrnDtreDVFVHqSAIaNnpkkavrSvgdGevVeFdvviGeKG  123 (U49120)
```

Figure 2. Cold Shock Domain in Plant Proteins. The entire CSPA sequence was used for a BLAST search but only the region of sequence with significant alignment is shown. E.c., Escherichia coli; N.s., Nicotiana sylvestris; A.t., Arabidopsis thaliana; D.m., Drosophila melanogaster. Numbers on the left and right of sequence indicate position in protein. Uppercase letters are identical with CSPA. Bold letters indicate residues that are perfectly conserved between CSPA and the four plant sequences. The underlined sequences represent the RNP1 and RNP2 motifs. The Y-box protein sequence of Drosophila (Thieringer et al., 1997) was included to show the relative identity between CSPA, the plant sequence and that of non-plant. Accession numbers are shown on the right.

is considered to function as an RNA chaperone. Effficient processing, transport, translation and degradation of RNA require single-stranded conformations. Cold shock domain proteins bind to RNA in association with DEAD-box RNA helicases and are thought to maintain single-stranded conformations (Sommerville, 1999).

Cold shock domain proteins in plants have been noted (Kingsley and Palis, 1994), but have not been studied in any detail. CSPA was used to do a BLAST search of the Arabidopsis database and all available plant sequences to determine whether proteins containing the cold shock domain were common in plants. Four sequences were found and the regions with the highest sequence identity were found to coincide with the cold-shock domain (Figure 2.). In fact, the result of a BLAST search of the GenBank database revealed that the four best matches to CSPA of all eukaryotes were these plant sequences. All four are glycine-rich proteins of unknown function that have not been extensively studied (de Oliveira et al., 1990; Obokata et al., 1991). Unfortunately, none have been examined for low temperature induction at this time or nucleic acid binding characteristics. Homologs to other Y-box related proteins, in particular the CCAAT-box binding complex of yeast and mammalian NF-Y, CBF or CP1 equivalents have been identified in Arabidopsis (Edwards et al., 1998). A rather large DEAD-box RNA helicase gene family is known in Arabidopsis (Aubourg et al., 1999). Similar to the cold shock domain proteins of plants, nothing is known about their role in transcriptional regulation, nucleic acid binding specificity or responsiveness to low temperature.

Molecular Convergence of Freezing Stress Responses with Other Osmotic Stresses

In the late 1980s and early 1990s, researchers were beginning to identify genes in plants that were differentially expressed in response to water deficits (Close et al., 1989), and low temperature (Hajela et al., 1990). While there was reason to view the two types of stresses as different physical processes, in the case of freezing stress, water is withdrawn from the cell and accretes to extracellular ice crystals during extracellular freezing (Levitt, 1972). This is the most common form of ice formation in freeze-tolerant plants. Naturally extracellular freezing results in the dehydration of the cells in frozen plants. Sequence analyses of major water stress induced genes and major low temperature induced genes revealed several families of genes containing a variety of repeating motifs with varying degrees of conservation. The genes were seemingly unique to plants and cyanobacteria (Close and Lammers, 1993), and had no known function. Many of the different classes of these genes that belong to the different superfamilies were first described as part of the developmental program of seed maturation in plants (Baker et al., 1988) known as late embryogenesis abundant (LEA) proteins. Many were found to accumulate during development just prior to the time when the seed undergoes dehydration and enters the relatively abiotic stress resistant dormant state, the plant equivalent to the bacterial spore. Genes that were isolated from plants of all types, from herbaceous annuals to trees subjected to low temperature, frequently belonged to the dehydrin family (Close et al., 1989). Dehydrins are synonymous with LEA D-11 family of developmentally programmed seed proteins (Baker et al., 1988). These proteins can reach levels as high as 1% of the total soluble protein (Close et al., 1989) during water or low temperature stress. They are recognized by a highly conserved 15 amino acid Lys-rich sequence EKKGIMDKIKEKLPG known as the K segment (Close, 1996). The K segments vary in number from one to 11 or more, and modeling studies suggest they could form an amphipathic α-helix. The core of the K segment approximates a class A amphipathic helix which is found in apolipoproteins (Close, 1996). A second sequence common to these proteins is (V/T)DEYGNP known as the Y segment that has relatedness to a nucleotide binding site of chaperones but is not widely found in the cold-inducible forms. Outside of the conserved repeat motifs, there is little conservation of sequence from member to member or the apparent orthologs of different plants. The high content of Gly and other polar amino acids in these proteins is consistent with limited results that suggest they may have large amounts of unordered character (Close 1996). If these proteins do contain large amounts of unordered structure, then the lack of conservation among these proteins in different species can be reasonably reconciled as there would be little or no selective restraint on mutations in the unordered regions.

The cold inducible members of the WCS120 family of wheat, also dehydrins, were found to be localized in the cytoplasm and nucleus (Houde et al., 1995). They accumulated to about 1% of the soluble protein with an estimated concentration of greater than 1 µM. In addition, they were most abundant in vascular transition zones. The high levels in this part of the plant are consistent with the notion that in regions of water transport they play a role in minimizing the deleterious effects of freezing. Another study showed that the cold regulated gene wcor410 is associated with the development of freeze tolerance in a number of gramineae species (Danyluk et al., 1998). EM immunolocalization showed that the protein WCOR410 is asymmetrically distributed with close association with the plasma membrane, and fractionation studies suggest that it is a peripheral protein. It has been proposed that WCOR410 helps to prevent destabilization of the membrane during freeze induced dehydration. This is consistent with the hypothesis that these proteins might act to stabilize hydrophobic surfaces during times of low water activity (Close, 1997).

In perennial plants, the ability of shoot meristems to survive freeze/thaw stress in winter is essential for the continued growth from year to year. In birch apices, a single dehydrin was found in the cytoplasm when plants were grown at warm temperatures. During low temperature acclimation, two additional dehydrins accumulated in nuclei, protein bodies and amyloplasts (Rinne et al., 1999). In vitro studies with a partially purified dehydrin demonstrated an ability to preserve the activity of starch α-amylase at reduced water activity, and support the hypothesis that dehydrins rescue metabolic processes.

It has not yet been possible to directly test the role of dehydrins and other water-stress related genes in stress tolerance of plants by making knock-out mutants. In some cases, these proteins have been ectopically expressed individually in other plants and the results have been largely disappointing with respect to altered stress tolerance (Artus et al, 1996; Kaye et al.,1998). In one case, however, a tomato gene (Le25) was expressed in yeast and found to increase tolerance to high salt and freezing, but not to high osmolarities caused by the presence of sorbitol in the medium (Imai et al., 1996). This gene is expressed in tomato in response to water stress and ABA and during seed maturation, but has very little expression in response to LT. A similar approach expressing another member of the LEA class of proteins from wheat in yeast resulted in a reduction of the detrimental effects of low water potentials (Swire-Clark and Marcotte, 1999). Continued success of heterologous expression of plant genes in yeast could help to dissect the mechanisms of stress protection by the various types of stress proteins. This appears to be the case as expression of Group 2 and Group 3 LEA genes in yeast revealed a functional divergence (Zhang et al., 2000). The Group 3 LEA was better able to preserve growth during NaCl stress than that of the Group 2 LEA, yet both increased freezing tolerance.

Perception of Low Temperature

How low temperature is perceived is a question of central importance. The early search for the controlling agent in cold acclimation and dormancy in plants focussed on the phytohormone abscisic acid (ABA). P.F. Wareing's lab in 1963 isolated a compound from *Betula pubescens* that when applied

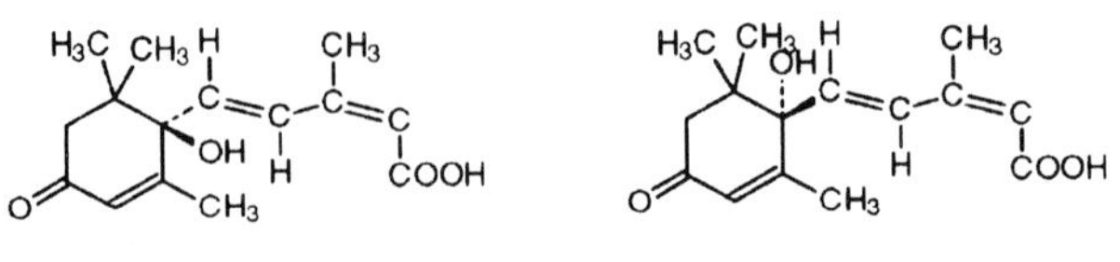

(S)-Abscisic Acid   (R)-Abscisic Acid

Figure 3. The Structure of Abscisic Acid

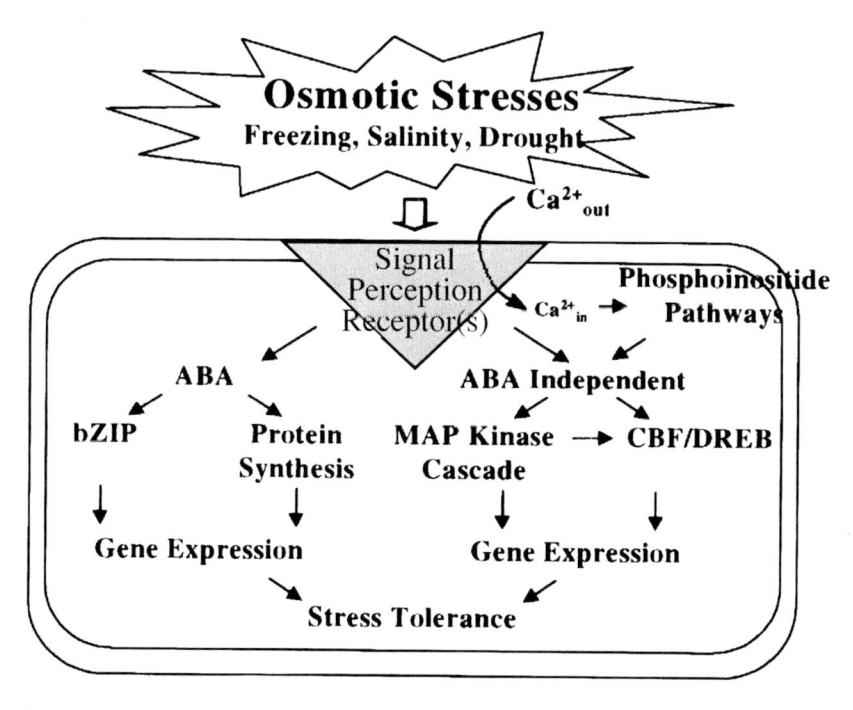

Figure 4. Outline of Known Major Osmotic Stress Responsive Signal Transduction Pathways in Plants

to the buds of growing seedlings would induce dormancy (Eagles and Wareing, 1963). The dormancy-promoting compound was given the name dormin. Its structure was determined and found to be identical with another compound abscisin (Ohkuma et al., 1963). The dormancy-promoting compound is known today as abscisic acid and this refers to the naturally occurring enantiomorph (S)-abscisic acid (Figure 3). The rate-limiting step in drought-induced ABA synthesis is catalyzed by 9-cis-epoxycarotenoid from a thylakoid-localized carotenoid substrate (Qin and Zeevaart, 1999). Based on RNA blot and ABA analyses, it has been proposed that ABA accumulation at low temperature may be more a function of dehydration than a low temperature response *per se* (Qin and Zeevaart, 1999).

When plants are exposed to low temperature, for many species, there is a transient rise in ABA content (Chen et al., 1983). This spike in ABA is associated with the onset of cold acclimation and the induction of freezing tolerance. In some cases, exogenous application of ABA to plantlets growing in vitro could increase freezing tolerance, but this response was never consistent. ABA is also strongly accumulated when plants experience a water deficit (Wright and Hiron, 1969). Interestingly, droughted plants often are found to be more resistant to freezing injury than when adequate water is available (Guy et al., 1992b).

In the early 1990s it became clear that many of the molecular responses of plants to drought, cold shock and cold acclimation included a subset of similar gene products (Close et al., 1989; Gilmour et al., 1992; Horvath et al., 1993; Nordin et al., 1993; Yamaguchi-Shinozaki et al., 1993). Given the similarity in these responses, it was suspected that perhaps ABA was a common agent that mediated responses to both types of stresses. Using mutants either unable to make ABA or unable to perceive the ABA signal, two independent groups showed that at least two separate signal transduction pathways existed (Figure 4), one that involved ABA and a second cold specific ABA-independent pathway for the genes that were responsive to both cold and water stress (Gilmour et al., 1991; Nordin et al., 1991).

Given the central role of calcium in cell signaling, several studies have focussed on its possible role in low temperature perception. One of the most direct approaches to address calcium used the calcium-sensitive luminescent protein, aequorin (Knight et al., 1991). The calcium-sensitive, aequorin, was introduced into tobacco plants using recombinant methods. Transformed plants constitutively producing the recombinant apoaequorin protein were used to report intracellular calcium levels following addition of coelenterazine. When plants were grown at 20°C and then the temperature decreased to 0°C - 5°C, a large transient influx of Ca^{2+} was observed. This contrasted with a heat shock where little influx was observed. The cold shock caused an immediate increase in cytosolic free calcium, which could be partially blocked by EGTA or lanthanum (Knight et al., 1996). Elevated cytosolic calcium seemed to be linked with the induction of cold-regulated gene expression because when calcium channels were inhibited by lanthanum or EGTA, so apparently was gene expression (Monroy and Dhindsa, 1995). At 25°C, the addition of a calcium ionophore stimulated the influx of labeled calcium and induced the expression of the cold regulated genes further supporting a role for Ca^{2+} as an early signaling agent in cold perception. The expression of two calcium dependent protein kinases was studied, and one (MSCK1) was strongly upregulated at 4°C within three hrs whereas two cold regulated genes examined required about 24 hrs for strong induction (Monroy and Dhindsa, 1995). This favored the idea that Ca^{2+} related events were rapidly activated in response to low temperature. Calcium dependent phosphorylation/dephosphorylation processes seem to be involved in activating the cold acclimation process. Consistent with Ca^{2+} channel blocker experiments, plants expressing CAX1, an Arabidopsis Ca^{2+}/H^+ antiporter, were hypersensitive to conditions of low Ca^{2+} or cold shock suggesting Ca^{2+} influx into the cytosol is an important factor in cold stress signaling and tolerance (Hirschi, 1999). Application of a protein kinase inhibitor blocked cold regulated gene expression and the induction of cold acclimation, while a protein phosphatase inhibitor, okadaic acid, stimulated the induction of the cold regulated gene cas15 (Monroy et al., 1998). Low temperature treatment caused a decrease in total protein phosphatase activity and a near complete inhibition of phosphatase 2A. Similarly, 2A activity could be decreased by calcium ionophore treatment. Recently a receptor-like protein kinase (RPK1) from Arabidopsis was found to be rapidly induced by several stresses (Hong et al., 1997). RPK1 is expressed in response to dehydration stress in several ABA metabolism and receptor mutants indicating it is regulated through an ABA independent mediated pathway. The subdomains of the C-terminal kinase domain are most closely similar to Ser/Thr kinases. It has a signal sequence consistent with an ER membrane localization along with a possible membrane spanning region.

Apparently there are well-conserved protein kinases that are essential for stress tolerance in both yeast and plant cells. One such kinase, At-Dbf2, of Arabidopsis when expressed in yeast significantly enhanced salt, drought, heat and cold tolerance (Lee et al., 1999). The Arabidopsis protein could functionally restore the yeast Dbf2 insertional mutant in strains transformed with At-Dbf2. The kinase is induced by several stresses including cold. Dbf2 in yeast is a component of a multi-subunit CCR4 general transcription complex. While At-Dbf2's role during cold stress remains undefined as part of a transcription complex, it may act to control the expression of genes that act to confer stress tolerance (Lee et al., 1999).

A well characterized effector of Ca^{2+} signaling in eukaryotes is a Ser/Thr specific protein phosphatase, calcineurin. This phosphatase has a regulatory (B) and a catalytic (C) subunit. Calcium activates the phosphatase via a Ca^{2+}-stimulated calmodulin interaction and acts as an effector of Ca^{2+} signaling. A calcineurin B-like protein, AtCBL1, has been identified in Arabidopsis that is strongly induced by drought, cold and wounding (Kudla et al., 1999). The limited set of stresses that activate expression of AtCBL1 indicates a specific function in the cellular responses to cold drought and wounding. However, what proteins are substrates and what cellular processes are regulated by the phosphatase are not yet known.

Inhibitors of the phosphoinositide signalling response decreased the magnitude of the cold shock induced Ca^{2+} influx suggesting a possible role in the calcium influx response (Knight et al., 1996). Targeting aequorin to the cytosolic face of the tonoplast membrane (a large calcium storage site inside the plant cell) showed that in combination with IP_3 metabolic inhibitors that at least part of the calcium influx could be attributed to the phosphoinsitide pathway and from vacuolar calcium stores. If cold is perceived or transduced by a calcium influx mechanism, then the signalling pathway needs only to be a brief spike in calcium levels to trigger the continuous expression of cold regulated genes.

Another approach to unraveling the signal transduction pathway(s) is to construct a reporter system using a chimeric synthetic gene consisting of a stress responsive promoter and the coding sequence for an enzyme whose activity is easily detectable using a non-destructive assay. This is precisely the approach that has been taken by Zhu and his colleagues. The *LUC* coding sequence was placed under the control of the stress-responsive promoter *rd29A* and transformed into Arabidopsis (Ishitani et al., 1997). Homozygous lines expressing the LUC gene were mutagenized with EMS and 500 – 1000 seed were grown on a defined medium. Following a variety of stress treatments, the plantlets were imaged for light production with luciferin. With this system, it was possible to identify several mutants in the sensing, signal transduction pathways. Some mutants were found to have constitutive expression *cos*, some gave low expression *los*, and some gave high expression *hos*. Fourteen different classes of mutants were identified based on the responses to one or a combination of stresses and ABA signals. Based on the analyses of the various mutant classes, a model for signal transduction was developed. There appears to be a cold mediated pathway that interfaces with ABA as well as a drought/salinity stress pathway that also interfaces with ABA. The cold and drought/salinity pathways converge into one transduction path that at some point converges with the ABA mediated pathway. Additional studies have yielded evidence for both positive and negative regulatory interactions among stress factors in the influence on gene expression (Xiong et al., 1999). Low temperature has a negative influence on osmotic or in combination with ABA mediated responses, but is additive with ABA application only. ABA responses were amplified by low temperature. Using transient expression assays with a similar LUC reporter gene and the regulatory region of the wcs120 gene, Ouellet and collegues (Ouellet et al, 1998) found evidence for the existence of both negative and positive regulatory mechanisms operating in wheat. The promoter region conferred cold regulation in both freezing tolerant and freezing sensitive monocots and dicots indicating that transcriptional activation machinery responsive to cold is widely present in plants.

Some of the signalling mutants show expression patterns that are useful in unraveling the signalling pathways. The *hos1-1* mutant results in the specific superinduction of several cold regulated genes. Since these genes are not superinduced in response to other normally inductive stresses, the HOS1 locus appears to be a negative regulator of cold signal transduction and perhaps a positive regulator for other stress signal pathways (Ishitani et al., 1998). The *hos1-1* plants are dwarfed compared to wild-type and flower early. *HOS2* mutants have enhanced expression of the *rd29A* gene and other stress responsive genes at low temperature. Enhanced expression does not occur in response to osmotic stress or ABA. *Hos2-1* is less freezing tolerant than wild-type, but another low temperature requiring process, vernalization, is unaffected in the mutant. *HOS2* appears to be a negative regulator of low temperature signal transduction for cold acclimation (Xiong et al., 1999).

Still another approach to unravel how low temperature is perceived and the signal transduced has been to look specifically for genes that are known to function in signal transduction pathways. Genes encoding two-component response regulator-like proteins were found to be induced by low temperature and other osmotic stresses in Arabidopsis (Urao et al., 1998). Presumably their function is to transduce extracellular signals into the interior of the cell. The two-component system consists of two proteins, a histidine kinase sensor and a response regulator. *E. coli* has scores of the kinase-regulator pairs. A number of histidine kinases have been cloned from plants showing that the two-component system is not confined to prokaryotes, however, few response regulators are known. The four described by Urao have the highly conserved Asp in the receiver domain.

Many stress and extracellular signals are mediated by mitogen-activated protein (MAP) kinase cacades in animals and yeast. When alfalfa plants were exposed to low temperature or drought, $p44^{MMK4}$, a MAP kinase, was transiently activated which was correlated with a decrease in electrophoretic mobility consistent with activation by phosphorylation (Jonak et al., 1996). The steady-state mRNA levels were also rapidly increased upon low temperature exposure, but the protein levels remained constant indicating that the kinase activation was a posttranslational activation process. When ABA was applied to alfalfa plants, p44^{MMK4} was not activated indicating that it acts on a signal transduction pathway that is independent of ABA. Similarly, several cDNAs for stress responsive genes from Arabidopsis were isolated that have homology to MAP kinase, and a MAPKK kinase, the latter having homology to the Byr2 from *Schizosaccharomyces pombe*, and Ste11 and Bck1 from *S. cerevisiae* (Mizoguchi et al., 1996). One of the cDNAs, ATMEKK1, can replace Ste11 in the response to mating pheromone in yeast. A third kinase related to the ribosomal S6 kinase was also studied. The mRNAs for all three kinases were strongly increased in response to cold and a variety of other stresses. The results are consistent with a role of the MAP kinase cascade as a transduction pathway for extracellular signals to function in transducing signals in response to cold stress.

Functional analysis of a MAP kinase cascade by oxidative stress has demonstrated activation of a cascade response by H_2O_2 involving the ANPl, a MAPKKK (Kovtun et al., 2000). Interestingly, ANPl appears to initiate the phosphorylation cascade. When ANPl is constitutively expressed transgenic tobacco exhibits general stress tolerance including enhanced freezing tolerance (Kovtun et al., 2000).

Similarly, two rare cold inducible mRNAs were cloned from Arabidopsis and sequence comparisons indicated that they belong to a family of proteins designated as 14-3-3 (Jarillo et al., 1994). These proteins participate in protein phosphorylation and protein-protein interactions by getting proteins together, a process that seems to be particularly important in signal transduction pathways. Another gene, (MP2C), encoding a functional protein phosphatase of the 2C type from alfalfa could target Ste11 of yeast and is therefore a MAPK kinase kinase, and a central regulator of osmosensing in yeast (Meskiene et al., 1998). MP2C may be a negative regulator of stress-activated MAPK activated by cold and other stresses and stimuli. Activation of stress-activated pathway involves posttranslational mechanisms, and negative regulation requires transcription and translation. Studies with the Wcs120 gene that is strongly expressed during cold acclimation, suggested multiple DNA-binding proteins bind to several elements in the promoter (Vazquez-Tello et al., 1998). Surprisingly, no DNA-binding activity was seen in extracts from low temperature treated plants. If such extracts were subject to dephosphorylation in vitro, DNA-binding activity was restored. Okadaic acid stimulated the accumulation of the Wcs120 family of proteins, thus it appears that protein phosphatases function to negatively repress expression of the Wcs120 family. Protein blot analyses indicated that an apparent protein kinase C homolog was selectively accumulated in the nucleus at low temperature. Thus it appears that the Wcs120 family, and perhaps many other cold responsive genes are regulated by factors whose activities are influenced by phosphorylation/dephosphorylation.

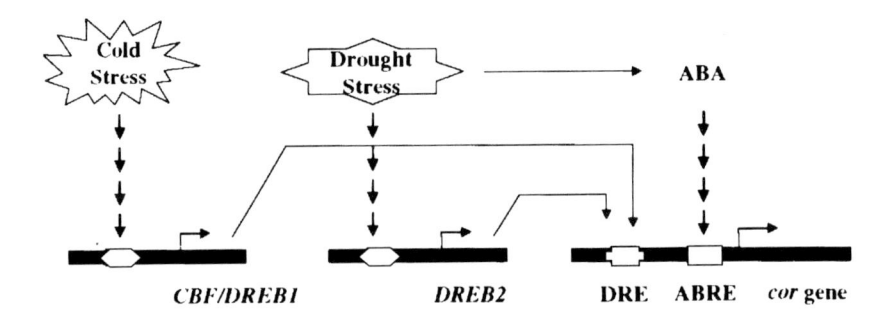

Figure 5. Transcriptional Activation Factors that Influence Cold Regulated Gene Expression

Transcriptional Activation

Promoter analyses by two different laboratories of cold regulated homolog genes in Arabidopsis and *Brassica napus* identified an element (TGGCCGAC) that was important in transcriptional activation at low temperature (Baker et al., 1994; White et al., 1994) and given the name C-repeat. At the about the same time, another group working on drought stress identified a novel *cis*-acting dehydration-responsive element (DRE) in a different set of genes. This latter work with Arabidopsis, used the tandem linked genes for *rd29A* and *rd29B* which are differentially regulated such that *rd29A* is cold, drought and salt induced while *rd29B* is not induced by cold (Yamaguchi-Shinozaki and Shinozaki, 1994). Deletion studies with *rd29A* revealed that a 9 bp element (TACCGACAT) was the DRE. Unexpectedly it contained at its center a 5 bp sequence that was also present in the element involved in cold regulation (see underlined bases). Comparison of promoter sequences of many cold regulated genes showed high conservation of a five bp motif CCGAC and further supported the identity of this sequence in low temperature regulation. When mutations were introduced into this 5-bp motif cold-regulation was eliminated confirming it as a *cis*-element in cold regulation (Jiang et al., 1996). Analyses of the promoter regions of blt4.9, a cold regulated member of non-specific lipid transfer protein family from barley indicated the presence of a putative binding site for a low temperature responsive transcription factor that consists of the sequence CCGAAA (Dunn et al., 1998). A sequence CCGAC in the promoter region similar to the C-repeat did not appear to bind nuclear proteins present in cold acclimated plants.

Once a *cis*-element involved in cold regulation was identified, the next step was to isolate the transcription factor that bound to the C-repeat. The first transcription factor involved in cold regulated gene expression in plants to be cloned was CBF1 (Stockinger et al., 1997). CBF1 contains nuclear localization sequence, acidic activation domain and DNA binding domain motif found in APETALA2 and other plant transcription factors. It binds to the C-repeat/DRE element known to be important in cold regulated genes. CBF1 was shown to be a functional transcription factor in vivo by its ability to activate reporter genes in yeast that contain the C-repeat/DRE in the promoter region. CBF1 is a low copy number transcriptional activator. In yeast CBF1 needed adaptor proteins for optimal function (Stockinger et al., 1997). Because CBF1 mRNA levels were not drastically changed by low temperature or water stress, it was suggested that perhaps CBF1 was activated by stress before it could be an activator of cor gene expression.

The DRE plays a role in responses to water stress and cold (Figure 5). Two cDNAs for DREB1A and DREB2A were obtained by a one-hybrid screen (Liu et al., 1998). The two cDNAs show significant sequence similarity only in the DNA binding domain. However, both bind to the same DRE element. DREB1A has two homologs 1B and 1C, and DREB1B is identical to CBF1. All show a similarity of about 86-87% at the amino acid level. One homolog of DREB2A was found (DREB2B) and showed about 54% similarity. DREB1A was strongly and rapidly cold induced but DREB2A was not. Instead DREB2A is strongly responsive to water and salt stress. DREB2A is responsive to ABA, but DREB1A is not. The three DREB1 genes are induced by cold, but not osmotic stress. A number of putative cis-elements were found in the promoters of the three genes (Shinwari et al., 1998), but not the C-repeat/DRE element (Gilmour et al., 1998). However, the pentamer CAGCC that is the reverse of the C-repeat/DRE is present in all three promoters (Medina et al.1999). None of the three transcription factor genes contain introns. Interestingly, they are arrayed in order B, A, C over 8.7 kb of chromosome IV (Gilmour et al., 1998; Shinwari et al., 1998; Medina et al., 1999).

Biophysical studies with recombinant CBF1 have revealed a reversible cold denaturation of the N-terminal and acidic regions of the protein at temperatures between 20°C and -5°C (Kanaya et al., 1999). CBF1 binds to C-repeat/DRE over this range of temperatures, but does not activate transcription at the higher temperatures. Instead it may act like a repressor at warm temperature, but become an activator when bound to the C-repeat/DRE at low temperature. The model for cold-regulated expression which seems to best fit available evidence is that another factor associates with CBF1 at warm temperature and acts to repress transcription. At low temperature CBF1 undergoes a temperature-induced change in structure that causes the cofactor to release and allows CBF1 to become a transcriptional activator.

Another line of evidence suggests that CBF may require activation to stimulate transcription of cold regulated genes containing C-repeat/DRE cis-elements. The sfr6 mutation of Arabidopsis confers sensitivity to freezing after cold acclimation. Genes regulated by CBF1 are not transcriptionally activated during cold acclimation while other cold regulated genes exhibit normal expression levels. The lack of expression by CBF1 regulated genes is tightly linked to sensitivity to freezing and implies that CBF1 must be activated before it can increase transcription (Knight et al., 1999).

Low temperature exposure of relatively cold sensitive corn and rice results in the induction of a leucine zipper DNA-binding factor that contains a bZIP motif (Aguan et al., 1993; Kusano et al., 1995). The gene is also strongly induced by salt stress and ABA, but not by water stress. When recombinant mLIP15 (the corn bZIP gene) was purified, it was shown to bind to the promoter of a wheat histone H3 gene. The binding could be diminished by adding excess amounts of a hexamer ACGTCA that is present in the histone promoter and in other G-box like elements.

Not all cold activation of transcription factors in plants is beneficial. Several tomato MADS-box genes showed greatly increased mRNA levels at low temperature and in situ hybridizations showed altered stage-specific expression for one in particular, *TM4* (Lozano et al., 1998). Low temperature caused disfunctional homeotic and meristematic alterations in flowers and reproductive whorls. Meristematic alterations occurred early in flower development while homeotic alterations occurred later during the formation of organ primordia. Floral meristem identity genes in plants are known. Many of the identity genes belong a to family of transcriptional activators having a conserved DNA-binding domain known as a MADS box. When tomato plants are exposed to low temperature, many flowers and fruits undergo an abnormal development. Some of the abnormalities include lack of stamen fusion, style splitting, stamens fused to carpels and carpelloid stamens.

Prospects for Genetic Modification of Temperature Stress Tolerance

Despite the apparent simplicity of the heat shock response, the complexity that underlies the basis of either acquired thermotolerance or cold acclimation in plants has not been overlooked. In both responses the changes in gene and protein expression are highly coordinated, and only over time has this fact become truly appreciated. Previously, it was thought that temperature stress tolerance might be altered by changing the expression of one or a very limited number of genes responsible for whatever tolerance mechanisms that contributed to enhanced survival. While in some systems there has been some success, most efforts to change tolerance by ectopic expression of one or two stress proteins has not been successful (Kaye et al., 1998). Such gene transfer efforts confirmed the long known fact that environmental stress tolerance of plants was a quantitative trait under the control of many genes (Frova and Sari-Gorla, 1994; Hayes et al., 1993; Pan et al., 1994).

Understanding the molecular basis for temperature stress responses and how changes in gene expression are regulated in response to stress offers an alternative to the single gene approach to stress tolerance enhancement. Transcription factors active in the heat shock response (Scharf et al., 1990) and for cold acclimation (Stockinger et al., 1997) of plants are now known. In both cases, the major transcription factors are activators of stress gene expression (Hubel et al., 1995; Stockinger et al., 1997), meaning that the functional transcriptional factor(s) stimulates the expression of the appropriate genes of the regulon. The key word here is <u>genes</u>. Both acquired thermotolerance, and most certainly, acclimation to cold are the consequence of the coordinate regulation of expression of many genes (Li et al., 1999). Therefore, what better way to invoke the stress response and potentially induce stress tolerance than to artificially modulate the abundance of functionally active transcriptional activators? By changing the expression of the activator it should be theoretically possible to alter the expression of all of the genes under its control that are involved in the stress response. Initial efforts aimed at changing the expression of the transcriptional activators involved in temperature stress responses have shown much promise in altering the temperature stress tolerance of plants. For example, overexpression of a heat shock transcription factor in Arabidopsis has resulted in the constitutive expression of the heat shock proteins and also enhanced thermotolerance (Lee et al., 1995; Prandl et al., 1998). Similarly, overexpression of CBF1, also in Arabidopsis, resulted in constitutive expression of cold regulated genes and increased survival of plants subjected to a freeze-thaw stress (Jaglo-Ottosen et al., 1998). More recently, the overexpression of DREB1Ab (identical to CBF1) and DREB1Ac in Arabidopsis was shown to not only enhance freeze-thaw stress tolerance, but also improved drought tolerance as well (Liu et al., 1998). While these early efforts have demonstrated the potential to engineer improved temperature stress tolerance, refinements will be necessary to eliminate the negative side effects of the constitutive expression of stress proteins whose continuous overexpression in the absence of stress appears to be counter productive for maximum growth and yield (Kasuga et al., 1999).

References

Aguan, K., Sugawara, K., Suzuki, N., and Kusano, T. 1993. Low-temperature-dependent expression of a rice gene encoding a protein with a leucine-zipper motif. Mol. Gen. Genet. 240: 1-8.

Appert, C., Logemann, E., Hahlbrock, K., Schmid, J., and Amrhein, N. 1994. Structural and catalytic properties of the four phenylalanine ammonia-lyase isoenzymes from parsley (*Petroselinum crispum* Nym.). Eur. J. Biochem. 225: 491-499.

Alschuler, M., and Mascarenhas, J.P. 1982. Heat shock proteins and effects of heat shock in plants. Plant Mol. Biol. 1: 103-115.

Antikainen, M., Griffith, M., Zhang, J., Hon, W.C., Yang, D.S.C., and Pihakaski-Maunsbach, K. 1996. Immunolocalization of antifreeze proteins in winter rye leaves, crowns, and roots by tissue printing. Plant Physiol. 110: 845-857.

Antikainen, M., and Griffith, M. 1997. Antifreeze protein accumulation in freezing-tolerant cereals. Physiol. Plant. 99: 423-432.

Artus, N.N., Uemura, M., Steponkus, P.L., Gilmour, S.J., Lin, C., and Thomashow, M.F. 1996. Constitutive expression of the cold-regulated *Arabidopsis thaliana COR15a* gene affects both chloroplast and protoplast freezing tolerance. Proc. Natl. Acad. Sci. USA. 93: 13404-13409.

Aubourg, S., Kreis, M., and Lecharny, A. 1999. The DEAD box RNA helicase family in *Arabidopsis thaliana.* Nucl. Acids Res. 27: 628-636.

Baker, J., Steele, C., and Dure, L. 1988. Sequence and characterization of 6 LEA proteins and their genes from cotton. Plant Mol. Biol. 11: 277-291.

Baker, S.S., Wilhelm, K.S., and Thomashow, M.F. 1994. The 5'-region of

Arabidopsis thaliana cor15a has *cis*-acting elements that confer cold-, drought- and ABA-regulated gene expression. Plant Mol. Biol. 24: 701-713.

Bodner, M., and Larcher, W. 1987. Chilling susceptibility of different organs and tissues of *Saintpaulia ionantha* and *Coffea arabica*. Angew. Botanik. 61: 225-242.

Burcham, T.S., Osuga, D.T., Yeh, Y., and Feeney, R.E. 1986. A kinetic description of antifreeze glycoprotein activity. J. Biol. Chem. 261: 6390-6397.

Carystinos, G.D., MacDonald, H.R., Monroy, A.F., Dhindsa, R.S., and Poole, R.J. 1995. Vacuolar H^+-translocating pyrophosphatase is induced by anoxia or chilling in seedlings of rice. Plant Physiol. 108: 641-649.

Chen, H.H., Li, P.H., and Brenner, M.L. 1983. Involvement of abscisic-acid in potato cold-acclimation. Plant Physiol. 71: 362-365.

Chun, J.U., Yu, X.M., and Griffith, M. 1998. Genetic studies of antifreeze proteins and their correlation with winter survival in wheat. Euphytica. 102: 219-226.

Close, T.J. 1996. Dehydrins: Emergence of a biochemical role of a family of plant dehydration proteins. Physiol. Plant. 97: 795-803.

Close, T.J. 1997. Dehydrins: A commonalty in the response of plants to dehydration and low temperature. Physiol. Plant. 100: 291-296.

Close, T.J., Kortt, A.A., and Chandler, P.M. 1989. A cDNA-based comparison of dehydration-induced proteins (dehydrins) in barley and corn. Plant Mol. Biol. 13: 95-108.

Close, T.J., and Lammers, P.J. 1993. An osmotic-stress protein of cyanobacteria is immunologically related to plant dehydrins. Plant Physiol. 101: 773-779.

Danyluk, J., Perron, A., Houde, M., Limin, A., Fowler, B., Benhamou, N., and Sarhan, F. 1998. Accumulation of an acidic dehydrin in the vicinity of the plasma membrane during cold acclimation of wheat. Plant Cell. 10: 623-638.

Didier, D.K., Schiffenbauer, J., Woulfe, S.L., Zacheis, M., and Schwartz, B.D. 1988. Characterization of the cDNA encoding a protein binding to the major histocompatibility complex class II Y box. Proc. Natl. Acad. Sci. USA. 85: 7322-7326.

De Oliveira, D.E., Seurinck, J., Inze, D., Van Montagu, M., and Botterman, J. 1990. Differential expression of five Arabidopsis genes encoding glycine-rich proteins. Plant Cell. 2: 427-436.

DeVries, A.L. 1971. Glycoproteins as biological antifreeze agents in antarctic fishes. Science. 172: 1152-1155.

Duman, J.G., and Olsen, T.M. 1993. Thermal hysteresis protein-activity in bacteria, fungi, and phylogenetically diverse plants. Cryobiol. 30: 322-328.

Dunn, M.A., White, A.J., Vural, S., and Hughes, M.A. 1998. Identification of promoter elements in a low-temperature-responsive gene (blt4.9) from barley (*Hordeum vulgare* L.). Plant Mol. Biol. 38: 551-564.

Eagles, C.F., and Wareing, P.F. 1963. Experimental induction of dormancy in *Betula pubescens*. Nature. 199: 874-875.

Edwards, D., Murray, J.A.H., and Smith, A.G. 1998. Multiple genes encoding the conserved CCAAT-box transcription factor complex are expressed in Arabidopsis. Plant Physiol. 117: 1015-1022.

Fowler, D.B., and Gusta, L.V. 1979. Selection for winterhardiness in wheat. I. Identification of genotypic variability. Crop Sci. 19: 769-772.

Frova, C., and Sari-Gorla, M. 1994. Quantitative trait loci (QTLs) for pollen thermotolerance detected in maize. Mol. Gen. Genet. 245: 424-430.

Gilmour, S.J., and Thomashow, M.F. 1991. Cold acclimation and cold-regulated gene expression in ABA mutants of Arabidopsis thaliana. Plant Mol. Biol. 17: 1233-1240.

Gilmour, S.J., Artus, N.N., and Thomashow, M.F. 1992. cDNA sequence analysis and expression of two cold-regulated genes of Arabidopsis thaliana. Plant Mol. Biol. 18: 13-21.

Gilmour, S.J., Zarka, D.G., Stockinger, E.J., Salazar, M.P., Houghton, J.M., and Thomashow, M.F. 1998. Low temperature regulation of the Arabidopsis CBF family of AP2 transcriptional activators as an early step in cold-induced COR gene expression. Plant J. 16: 433-442.

Goldstein, J., Pollitt, N.S., and Inouye, M. 1990. Major cold shock protein of Escherichia coli. Proc. Natl. Acad. Sci. USA. 87: 283-287.

Graumann, P.L., and Marahiel, M.A. 1998. A superfamily of proteins that contain the cold-shock domain. Trends Biochem. Sci. 23: 286-290.

Griffith, M., Ala, P., Yang, D.S.C., Hon, W.C., and Moffatt, B.A. 1992. Antifreeze protein produced endogenously in winter rye leaves. Plant Physiol. 100: 593-596.

Griffith, M., Antikainen, M., Hon, W.C., Pihakaski-Maunsbach, K., Yu, X.M., Chun, J.U., and Yang, D.S.C. 1997. Antifreeze proteins in winter rye. Physiol. Plant. 100: 327-332.

Guy, C.L., and Carter, J.V. 1984. Characterization of partially purified glutathione-reductase from cold-hardened and nonhardened spinach leaf tissue. Cryobiol. 21: 454-464.

Guy, C.L., Niemi, K.J., and Brambl, R. 1985. Altered gene expression during cold acclimation of spinach. Proc. Natl. Acad. Sci. USA. 82: 3673-3677.

Guy, C.L., and Haskell, D. 1987. Induction of freezing tolerance in spinach is associated with the synthesis of cold acclimation induced proteins. Plant Physiol. 84: 872-878.

Guy, C.L., Huber, J.L.A., and Huber, S.C. 1992a. Sucrose phosphate synthase and sucrose accumulation at low temperature. Plant Physiol. 100: 502-508.

Guy, C.L., Haskell, D., Neven, L., Klein, P., and Smelser, C. 1992b. Hydration-state-responsive proteins link cold and drought stress in spinach. Planta. 188: 265-270.

Guy, C., Haskell, D., Li, Q.-B., and Zhang, C. 1997. Molecular chaperones: Do they have a role in cold stress responses of plants?, In: Fifth International Plant Cold Hardiness Seminar. P. Li and T. Chen, eds. Plenum Press, New York. p. 109-129.

Hajela, R.K., Horvath, D.P., Gilmour, S.J., and Thomashow, M.F. 1990. Molecular-cloning and expression of cor (Cold-regulated) genes in

Arabidopsis thaliana. Plant Physiol. 93: 1246-1252.

Hayes, P.M., Blake, T., Chen, T.H.H., Tragoonrung, S., Chen, F., Pan, A., and Liu, B. 1993. Quantitative trait loci on barley (*Hordeum vulgare*) chromosome 7 associated with components of winterhardiness. Genome. 36: 66-71.

Hirschi, K.D. 1999. Expression of Arabidopsis *CAX1* in tobacco: Altered calcium homeostasis and increased stress sensitivity. Plant Cell 11: 2113-2122.

Holaday, A.S., Martindale, W., Alred, R., Brooks, A.L., and Leegood, R.C. 1992. Changes in the activities of enzymes of carbon metabolism in leaves during exposure of plants to low temperature. Plant Physiol. 98: 1105-1114.

Hon, W.C., Griffith, M., Chong, P.L., and Yang, D.S.C. 1994. Extraction and isolation of antifreeze proteins from winter rye (*Secale cereale* L) leaves. Plant Physiol. 104: 971-980.

Hon, W.C., Griffith, M., Mlynarz, A., Kwok, Y.C., and Yang, D.S.C. 1995. Antifreeze proteins in winter rye are similar to pathogenesis-related proteins. Plant Physiol. 109: 879-889.

Hong, S.W., Jon, J.H., Kwak, J.M., and Nam, H.G. 1997. Identification of a receptor-like protein kinase gene rapidly induced by abscisic acid, dehydration, high salt, and cold treatments in *Arabidopsis thaliana*. Plant Physiol. 113: 1203-1212.

Horvath, D.P., McLarney, B.K., and Thomashow, M.F. 1993. Regulation of *Arabidopsis thaliana* L. (Heyn) cor78 in response to low temperature. Plant Physiol. 103: 1047-1053.

Houde, M., Daniel, C., Lachapelle, M., Allard, F., Laliberte, S., and Sarhan, F. 1995. Immunolocalization of freezing-tolerance-associated proteins in the cytoplasm and nucleoplasm of wheat crown tissues. Plant J. 8: 583-593.

Hubel, A., Lee, J.H., Wu, C., and Schöffl, F. 1995. Arabidopsis heat shock factor is constitutively active in *Drosophila* and human cells. Mol. Gen. Genet. 248: 136-141.

Hughes, M.A., and Dunn, M.A. 1996. The molecular biology of plant acclimation to low temperature. J. Exp. Bot. 296: 291-305.

Hurry, V.M., Malmberg, G., Gardeström, P., and Öquist, G. 1994. Effects of a short-term shift to low temperature and of long-term cold hardening on photosynthesis and ribulose-1,5-bisphosphate carboxylase/oxygenase and sucrose phosphate synthase activity in leaves of winter rye (*Secale cereale* L.). Plant Physiol. 106: 983-990.

Imai, R., Chang, L., Ohta, A., Bray, E.A., and Takagi, M. 1996. A lea-class gene of tomato confers salt and freezing tolerance when expressed in *Saccharomyces cerevisiae*. Gene. 170: 243-248.

Ishitani, M., Xiong, L., Stevenson, B., and Zhu, J.K. 1997. Genetic analysis of osmotic and cold stress signal transduction in Arabidopsis: interactions and convergence of abscisic acid-dependent and abscisic acid-independent pathways. Plant Cell. 9: 1935-1949.

Ishitani, M., Xiong, L., Lee, H., Stevenson, B., and Zhu, J.K. 1998. HOS1, a genetic locus involved in cold-responsive gene expression in Arabidopsis.

Plant Cell. 10:1151-1161.

Ito, Y., Saisho, D., Nakazono, M., Tsutsumi, N., and Hirai, A. 1997. Transcript levels of tandem-arranged alternative oxidase genes in rice are increased by low temperature. Gene. 203: 121-129.

Jaglo-Ottosen, K.R., Gilmour, S.J., Zarka, D.G., Schabenberger, O., and Thomashow, M.F. 1998. Arabidopsis CBF1 overexpression induces COR genes and enhances freezing tolerance. Science. 280: 104-106.

Jarillo, J.A., Capel, J., Leyva, A., Martinez-Zapater, J.M., and Salinas, J. 1994. Two related low-temperature-inducible genes of Arabidopsis encode proteins showing high homology to 14-3-3 proteins, a family of putative kinase regulators. Plant Mol. Biol. 25: 693-704.

Jiang, C., Iu, B., and Singh, J. 1996. Requirement of a CCGAC cis-acting element for cold induction of the BN115 gene from winter *Brassica napus*. Plant Mol. Biol. 30: 679-684.

Jiang, W., Hou, Y., and Inouye, M. 1997. CspA, the major cold-shock protein of *Escherichia coli*, is an RNA chaperone. J. Biol. Chem. 272: 196-202.

Jonak, C., Kiegerl, S., Ligterink, W., Barker, P.J., Huskisson, N.S., and Hirt, H. 1996. Stress signaling in plants: A mitogen-activated protein kinase pathway is activated by cold and drought. Proc. Natl. Acad. Sci. USA. 93: 11274-11279.

Kanaya, E. Nakajima, N., Morikawa, K., Okada, K. Shimura, Y. 1999. Characterization of the transcriptional activator CBF1 from *Arahidopsis thaliana*. J. Biol. Chem. 274: 16068-16076.

Kaye, C., Neven, L., Hofig, A., Li, Q.B., Haskell, D., and Guy, C. 1998. Characterization of a gene for spinach CAP160 and expression of two spinach cold-acclimation proteins in tobacco. Plant Physiol. 116: 1367-1377.

Khan, R.A. 1976. Effect of high-temperature stress on the growth and seed characteristics of barley and cotton. Basic Life Sci. 8: 319-324.

Kidou, S., Umeda, M., Tsuge, T., Kato, A., and Uchimiya, H. 1994. Isolation and characterization of a rice cDNA similar to the S-phase-specific cyc07 gene. Plant Mol. Biol. 24: 545-547.

Kingsley, P.D., and Palis, J. 1994. GRP2 proteins contain both CCHC zinc fingers and a cold shock domain. Plant Cell. 6: 1522-123.

Knight, M.R., Campbell, A.K., Smith, S.M., and Trewavas, A.J. 1991. Transgenic plant aequorin reports the effects of touch and cold-shock and elicitors on cytoplasmic calcium. Nature. 352: 524-526.

Knight, H., Trewavas, A.J., and Knight, M.R. 1996. Cold calcium signaling in Arabidopsis involves two cellular pools and a change in calcium signature after acclimation. Plant Cell. 8: 489-503.

Knight, H. Veale, E.L., Warren, G.J., and Knight, M.R. 1999. The *sfr6* mutation in Arabidopsis suppresses low-temperature induction of genes dependent on the CRT/DRE sequence motif. Plant Cell 11: 875-886.

Kodama, H., Akagi, H., Kusumi, K., Fujimura, T., and Iba, K. 1997. Structure, chromosomal location and expression of a rice gene encoding the microsome omega-3 fatty acid desaturase. Plant Mol. Biol. 33: 493-502.

Koukalova, B., Kovarik, A., Fajkus, J., and Siroky, J. 1997. Chromatin fragmentation associated with apoptotic changes in tobacco cells exposed to cold stress. FEBS Lett. 414: 289-292.

Kovtun, Y., Chiu, W.-L., Tena, G., and Sheen, J. 2000. Functional analysis of oxidative stress-activated mitogen-activated protein kinase cascade in plants. Proc. Natl. Acad. Sci. USA. 97: 2940-2945.

Kudla, J., Xu, Q., Hater, K., Gruissem, W., and Luan, S. 1999. Genes for calcineurin B-like proteins in *Arabidopsis* are differentially regulated by stress signals. Proc. Natl. Acad. Sci. USA. 96: 4718-4723.

Kusano, T., Berberich, T., Harada, M., Suzuki, N., and Sugawara, K. 1995. A maize DNA-binding factor with a bZIP motif is induced by low temperature. Mol. Gen. Genet. 248: 507-517.

Lee, J.H., Hubel, A., and Schöffl, F. 1995. Derepression of the activity of genetically engineered heat shock factor causes constitutive synthesis of heat shock proteins and increased thermotolerance in transgenic Arabidopsis. Plant J. 8: 603-612.

Levitt, J. 1972. Responses of Plants to Environmental Stresses. Academic Press, New York. p. 697.

Li, Q.B., Haskell, D.W., and Guy, C.L. 1999. Coordinate and non-coordinate expression of the stress 70 family and other molecular chaperones at high and low temperature in spinach and tomato. Plant Mol. Biol. 39: 21-34.

Liu, Q., Kasuga, M., Sakuma, Y., Abe, H., Miura, S., Yamaguchi-Shinozaki, K., and Shinozaki, K. 1998. Two transcription factors, DREB1 and DREB2, with an EREBP/AP2 DNA binding domain separate two cellular signal transduction pathways in drought- and low-temperature-responsive gene expression, respectively, in Arabidopsis. Plant Cell. 10: 1391-1406.

Kasuga, M., Liu, Q., Miura, S., Yamaguchi-Shinozaki K., and Shinozaki, K. 1999. Improving plant drought, salt, and freezing tolerance by gene transfer of a single stress-inducible transcription factor. Nat. Biotechnol. 17: 287-291.

Lozano, R., Angosto, T., Gomez, P., Payan, C., Capel, J., Huijser, P., Salinas, J., and Martinez-Zapater, J.M. 1998. Tomato flower abnormalities induced by low temperatures are associated with changes of expression of MADS-Box genes. Plant Physiol. 117: 91-100.

Lundegårdh, H. 1954. Klima und Boden in ihrer Wirkung auf das Pflanzenleben, 4th ed. Fischer, Jena.

Medina, J., Bargues, M., Terol, J., Perez-Alonso, M., and Salinas, J. 1999. The Arabidopsis CBF gene family is composed of three genes encoding AP2 domain-containing proteins whose expression is regulated by low temperature but not by abscisic acid or dehydration. Plant Physiol. 119: 463-470.

Meskiene, I., Bogre, L., Glaser, W., Balog, J., Brandstotter, M., Zwerger, K., Ammerer, G., and Hirt, H. 1998. P2C, a plant protein phosphatase 2C, functions as a negative regulator of mitogen-activated protein kinase pathways in yeast and plants. Proc. Natl. Acad. Sci. USA. 95: 1938-1943.

Meyer, K. Keil, M., and Naldrett, M.J. 1 999. A leucine repeat protein of carrot that exhibits antifreeze activity. FEBS Lett. 477: 1 71-178.

Mineur, P., Jennane, A., Thiry, M., Deltour, R., and Goessens, G. 1998. Ultrastructural distribution of DNA within plant meristematic cell nucleoli during activation and the subsequent inactivation by a cold stress. J. Struct. Biol. 123: 199-210.

Mizoguchi, T., Irie, K., Hirayama, T., Hayashida, N., Yamaguchi-Shinozaki, K., Matsumoto, K., and Shinozaki, K. 1996. A gene encoding a mitogen-activated protein kinase kinase kinase is induced simultaneously with genes for a mitogen-activated protein kinase and an S6 ribosomal protein kinase by touch, cold, and water stress in *Arabidopsis thaliana*. Proc. Natl. Acad. Sci. USA. 93: 765-769.

Monroy, A.F., and Dhindsa, R.S. 1995. Low-temperature signal transduction: induction of cold acclimation-specific genes of alfalfa by calcium at 25 degrees C. Plant Cell. 7: 321-331.

Monroy, A.F., Sangwan, V., and Dhindsa, R.S. 1998. Low temperature signal transduction during cold acclimation: protein phosphatase 2A as an early target for cold-inactivation. Plant J. 13: 653-660.

Muench, D.G., Wu, Y., Zhang, Y., Li, X., Boston, R.S., and Okita, T.W. 1997. Molecular cloning, expression and subcellular localization of a BiP homolog from rice endosperm tissue. Plant Cell Physiol. 38: 404-412.

Murray, M.T. 1994. Nucleic acid-binding properties of the Xenopus oocyte Y box protein mRNP3+4. Biochem. 33: 13910-13917.

Neumann, D., Nover, L., Parthier, B., Rieger, R., Scharf, K.-D., Wollgiehn, R., and zur Nieden, U. 1989. Heat shock and other stress response systems of plants. Biol. Zent. 108: 1-155.

Nordin, K., Heino, P., and Palva, E.T. 1991. Separate signal pathways regulate the expression of a low-temperature-induced gene in *Arabidopsis thaliana* (L.) Heynh. Plant Mol. Biol. 16: 1061-1071.

Nordin, K., Vahala, T., and Palva, E.T. 1993. Differential expression of two related, low-temperature-induced genes in *Arabidopsis thaliana* (L.) Heynh. Plant Mol. Biol. 21: 641-653.

Nover, L., and Scharf, K.-D. 1997. Heat stress proteins and transcription factors. Cell. Mol. Life Sci. 53: 80-103.

Obokata, J., Ohme, M., and Hayashida, N. 1991. Nucleotide sequence of a cDNA clone encoding a putative glycine-rich protein of 19.7 kDa in *Nicotiana sylvestris*. Plant Mol. Biol. 17: 953-955.

Ohkuma, K., Lyon, J.L., Addicott, F.T., and Smith, O.E. 1963. Abscisin II, an abscission-accelerating substance from young cotton fruit. Science. 142 1592-1593.

Ouellet, F., Vazquez-Tello, A., and Sarhan, F. 1998. The wheat wcs120 promoter is cold-inducible in both monocotyledonous and dicotyledonous species. FEBS Lett. 423: 324-328.

Pan, A., Hayes, P.M., Chen, F., Chen, T.H.H., Blake, T., Wright, S., Karsai, I., and Bedo, Z. 1994. Genetic analysis of the components of winterhardiness in barley (*Hordeum vulgare* L.). Theoret. Appl. Genet. 89: 900-910.

Pareek, A., Singla, S.L., and Grover, A. 1995. Immunological evidence for accumulation of two high-molecular-weight (104 and 90 kDa) HSPs in response to different stresses in rice and in response to high temperature stress in diverse plant genera. Plant Mol. Biol. 29: 293-301.

Pisek, A., Larcher, W., Vegis, A., and Napp-Zin, K. 1973. The normal temperature range. In: Temperature and Life. H. Precht, J. Christopherson, H. Hensel, and W. Larcher, eds. Springer-Verlag, Berlin. p. 102-194.

Prandl, R., Hinderhofer, K., Eggers-Schumacher, G., and Schöffl, F. 1998. HSF3, a new heat shock factor from *Arabidopsis thaliana*, derepresses the heat shock response and confers thermotolerance when overexpressed in transgenic plants. Mol. Gen. Genet. 258: 269-278.

Qin, X., and Zeevaart, J.A.D. 1999. 9-cis-epoxycarotenoid cleavage reaction is the key regulatory step of abscisic acid biosynthesis in water-stressed bean. Proc. Natl. Acad. Sci. USA. 96: 15354-15361.

Sakai, A., and Yoshida, S. 1968. The role of sugar and related compounds in variations of freezing resistance. Cryobiol. 5: 160-174.

Scharf, K.D., Rose, S., Zott, W., Schöffl, F., and Nover, L. 1990. Three tomato genes code for heat stress transcription factors with a region of remarkable homology to the DNA-binding domain of the yeast HSF. EMBO J. 9: 4495-4501.

Scorza, R., Ashworth, E.N., Bell, R.L., and Lightner, G.W. 1983. Sampling for field evaluation of peach and nectarine flower bud survival *Prunus persica*, statistics, winter hardiness. J. Amer. Soc. Hort. Sci. 108: 747-750.

Shinwari, Z.K., Nakashima, K., Miura, S., Kasuga, M., Seki, M., Yamaguchi-Shinozaki, K., and Shinozaki, K. 1998. An Arabidopsis gene family encoding DRE/CRT binding proteins involved in low-temperature-responsive gene expression. Biochem. Biophys. Res. Commun. 250: 161-170.

Singer, M.A., and Lindquist, S. 1998. Multiple effects of trehalose on protein folding in vitro and in vivo. Mol. Cell. 1: 639-648.

Sommerville, J. 1999. Activities of cold-shock domain proteins in translation control. Bioessays. 21: 319-325.

Strand, A., Hurry, V., Henkes, S., Huner, N., Gustafsson, P., Gardestrom, P., and Stitt, M. 1999. Acclimation of Arabidopsis leaves developing at low temperatures. Increasing cytoplasmic volume accompanies increased activities of enzymes in the Calvin cycle and in the sucrose-biosynthesis pathway. Plant Physiol. 119: 1387-1398.

Stockinger, E.J., Gilmour, S.J., and Thomashow, M.F. 1997. *Arabidopsis thaliana* CBF1 encodes an AP2 domain-containing transcriptional activator that binds to the C-repeat/DRE, a *cis*-acting DNA regulatory element that stimulates transcription in response to low temperature and water deficit. Proc. Natl. Acad. Sci. USA. 94: 1035-1040.

Sukumaran, N.P., and Weiser, C.J. 1972. Freezing injury in potato leaves. Plant Physiol. 50: 564-567.

Swire-Clark, G.A., and Marcotte, W.R. 1999. The wheat LEA protein Em functions as an osmoprotective molecule in *Saccharomyces cerevisiae*. Plant Mol. Biol. 39: 117-128.

Tajima, K., Amemiya, A., and Kabaki, N. 1983. Physiological study of growth inhibition in rice plant as affected by low temperature. II. Physiological mechanism and varietal difference of chilling injury in rice plant. Bull. Natl. Inst. Agr. Sci. D34: 69-111.

Takahashi, R., Joshee, N., and Kitagawa, Y. 1994. Induction of chilling resistance by water stress, and cDNA sequence analysis and expression of water stress-regulated genes in rice. Plant Mol. Biol. 26: 339-352.

Taylor, A.O., Slack, C.R., and McPherson, H.G. 1974. Plants under climatic stress. VI. Chilling and light effects on photosynthetic enzymes of sorghum and maize. Plant Physiol. 54: 696-701.

Thieringer, H.A., Singh, K., Trivedi, H., and Inouye, M. 1997. Identification and developmental characterization of a novel Y-box protein from *Drosophila melanogaster*. Nucl. Acids Res. 25: 4764-4770.

Thomashow, M.F. 1998. Role of cold-responsive genes in plant freezing tolerance. Plant Physiol. 118: 1-8.

Thomashow, M.F. 1999. Plant cold acclimation: Freezing tolerance genes and regulatory mechanisms. Annu. Rev. Plant Physiol. Plant Mol. Bio. 50: 571-599.

Uemura, M., and Steponkus, P.L. 1994. A contrast of the plasma-membrane lipid-composition of oat and rye leaves in relation to freezing tolerance. Plant Physiol. 104: 479-496.

Uemura, M., Joseph, R.A., and Steponkus, P.L. 1995. Cold acclimation of *Arabidopsis thaliana*: Effect on plasma membrane lipid composition and freeze induced lesions. Plant Physiol. 109: 15-30.

Urao, T., Yakubov, B., Yamaguchi-Shinozaki, K., and Shinozaki, K. 1998. Stress-responsive expression of genes for two-component response regulator-like proteins in *Arabidopsis thaliana*. FEBS Lett. 427: 175-178.

Urrutia, M.E., Duman, J.G., and Knight, C.A. 1992. Plant thermal hysteresis proteins. Biochim. Biophys. Acta. 1121: 199-206.

Vazquez-Tello, A., Ouellet, F., and Sarhan, F. 1998. Low temperature-stimulated phosphorylation regulates the binding of nuclear factors to the promoter of Wcs120, a cold-specific gene in wheat. Mol. Gen. Genet. 257: 157-166.

Vierling, E. 1991. The roles of heat shock proteins in plants. Ann. Rev. Plant Physiol. Plant Mol. Biol. 42: 579-620.

Wang, Q., and Fang, R. 1996. Structure and expression of a rice hsp70 gene. Sci. China C. Life Sci. 39: 291-299.

White, T.C., Simmonds, D., Donaldson, P., and Singh, J. 1994. Regulation of BN115, a low-temperature-responsive gene from winter *Brassica napus*. Plant Physiol. 106: 917-928.

Wolber, P.K., Deininger, C.A., Southworth, M.W., Vandekerckhove, J., van Montagu, M., and Warren, G.J. 1986. Identification and purification of a bacterial ice-nucleation protein. Proc. Natl. Acad. Sci. USA. 83: 7256-7260.

Worrall, D., Elias, L., Ashford, D., Smallwood, M., Sidebottom, C., Lillford, P., Telford, J., Holt, C., and Bowles, D. 1998. A carrot leucine-rich-repeat protein that inhibits ice recrystallization. Science. 282: 115-7.

Wright, S.T.C., and Hiron, R.W.P. 1969. Abscisic acid, the growth inhibitor induced in detached wheat leaves by a period of wilting. Nature. 224: 719-720.

Xiong, L., Ishitani, M., and Zhu, J.K. 1999. Interaction of osmotic stress, temperature, and abscisic acid in the regulation of gene expression in Arabidopsis. Plant Physiol. 119: 205-212.

Xu, H., Griffith, M., Patten, C.L., and Glick B.R. 1998. Isolation and characterization of an antifreeze protein with ice nucleation activity from the plant growth promoting rhizobacterium *Pseudomonas putida* GR12-2. Can. J. Microbiol. 44: 64-73.

Yamaguchi-Shinozaki, K., and Shinozaki, K. 1993. The plant hormone abscisic acid mediates the drought-induced expression but not the seed-specific expression of rd22, a gene responsive to dehydration stress in *Arabidopsis thaliana*. Mol. Gen. Genet. 238: 17-25.

Yamaguchi-Shinozaki, K., and Shinozaki, K. 1994. A novel cis-acting element in an Arabidopsis gene is involved in responsiveness to drought, low-temperature, or high-salt stress. Plant Cell. 6: 251-264.

Yeh, C.H., Chang, P.F., Yeh, K.W., Lin, W.C., Chen, Y.M., and Lin, C.Y. 1997. Expression of a gene encoding a 16.9-kDa heat-shock protein, Oshsp16.9, in *Escherichia coli* enhances thermotolerance. Proc. Natl. Acad. Sci. USA. 94: 10967-10972.

Yelenosky, G., and Guy, C.L. 1989. Freezing tolerance of citrus, spinach, and petunia leaf tissue. Plant Physiol. 89: 444-451.

Zhang, L., Ohta, A., Takagi, M., and Imai, R. 2000. Expression of plant Group 2 and Group 3 *lea* genes in *Saccharomyces cerevisiae* revealed functional divergence among LEA proteins. J. Biochem. 127: 611-616.

7

Cold Shock Response in Mammalian Cells

Jun Fujita

Department of Clinical Molecular Biology,
Faculty of Medicine, Kyoto University, Kyoto, Japan

Abstract

Compared to bacteria and plants, the cold shock response has attracted little attention in mammals except in some areas such as adaptive thermogenesis, cold tolerance, storage of cells and organs, and recently, treatment of brain damage and protein production. At the cellular level, some responses of mammalian cells are similar to microorganisms; cold stress changes the lipid composition of cellular membranes, and suppresses the rate of protein synthesis and cell proliferation. Although previous studies have mostly dealt with temperatures below 20°C, mild hypothermia (32°C) can change the cell's response to subsequent stresses as exemplified by APG-1, a member of the HSP110 family. Furthermore, 32°C induces expression of CIRP (cold-inducible RNA-binding protein), the first cold shock protein identified in mammalian cells, without recovery at 37°C. Remniscent of HSP, CIRP is also expressed at 37°C and developmentary regulated, possibly working as an RNA chaperone. Mammalian cells are metabolically active at 32°C, and cells may survive and respond to stresses with different strategies from those at 37°C. Cellular and molecular biology of mammalian cells at 32°C is a new area expected to have considerable implications for medical sciences and possibly biotechnology.

Introduction

In response to the ambient temperature shift, organisms change various physiological functions. Elevated temperatures was first dicovered to induce a set of proteins, heat shock proteins (HSPs) in *Drosophila* (Tissieres et al., 1974). Subsequently, HSPs were discovered in most prokaryotes and eukaryotes (Lindquist and Craig, 1988). Although the optimum temperature range of HSP induction varies considerably with the organism, it seems to be related to the physiological range of supraoptimal temperatures within which active adaptation is observed. For example this is around 40-50°C for birds and mammals, 35-37°C for yeasts and 35-40°C for plants. However, the optimum can vary between different cell types of a single organism, and between individual HSPs from even one cell type (Burdon, 1987). HSPs are also present in cells at normal temperatures and are now recognized as molecular chaperones, assisting in the folding/unfolding, assembly/disassembly and transport of various proteins (Morimoto, 1994).

Less is known about the cold shock responses. In microorganisms, cold stress induces the synthesis of several cold-shock proteins (Jones and Inouye, 1994). A variety of plant genes are known to be induced by cold stress, and are thought to be involved in the stress tolerance of the plant (Shinozaki and Yamaguchi-Shinozaki, 1996; Hughes *et al.*, 1999). The response to cold stress in mammals, however, has attracted little attention except in a few areas such as adaptive thermogenesis, cold tolerance, and storage of cells and organs. Recently, hypothermia is gaining popularity in emergency clinics as a novel therapeutic modality for brain damages. In addition, low temperature cultivation has been dicussed as a method to improve heterologous protein production in mammalian cells (Giard *et al.*, 1982).

Adaptive thermogenesis refers to a component of energy expenditure, which is separable from physical activity. It can be elevated in response to changing environmental conditions, most notably cold exposure and overfeeding. There has been considerable interest in this subject because of potential roles in obesity. Cold is sensed in the central nervous system and "cold-induced" expression of several genes, *e.g.* uncoupling protein *(UCP)-1* and *PGC-1*, are mediated by increased "sympathetic" output to peripheral tissues (Puigserver *et al.*, 1998). Even the induction of HSPs in brown adipose tissue in mice exposed to cold ambient temperature has been shown to be mediated by norepinephrin released in response to cold (Matz *et al.*, 1995).

In clinics, hypothermia has been employed in heart and brain surgery and in the preservation of organs to be used for transplantation. During cardiac surgery, protection against myocardial ischemia is attained through reduction of oxygen demand by minimizing electromechanical activity with potassium arrest and by reducing basal metabolic rate with hypothermia (Mauney and Kron, 1995). Hypoxic brain damage initiates several metabolic processes that can exacerbate the injury. Mild hypothermia is supposed to limit some of these deleterious metabolic responses, *e.g.* by altering the neurotransmitter release, attenuating energy depletion, decreasing radical oxygen species production and reducing neuronal death (Busto *et al.*, 1987; Bertman *et al.*, 1981; Connolly *et al.*, 1962). Clinically, beneficial effects of mild hypothermia (32-33°C) have been observed in patients with severe traumatic brain injury, elevated intracranial pressure and a critically low cerebral perfusion pressure (Marion *et al.*, 1997; Hayashi, 1998; Wassmann *et al.*, 1998). Elucidating the mechanisms of mammalian hibernation may be of use in developing clinically effective measures that prevent and/or cure the brain ischemia and damages. In fact, hippocampal slices from hibernating ground squirrel show increased tolerance to a superimposed hypoxia even at 36°C (Frerichs and Hallenbeck, 1998). However, the cellular or molecular mechanisms that trigger and maintain this adaptation remain unknown.

Until recently, the cold stress response has mainly been analyzed after exposing rodents and humans to cold ambient temperatures. It should be remembered, however, that exposing whole non-hibernating animals to cold may not lower the temperature of the tissues in which the "cold-induced" expression of genes are to be examined. For example, in one study incubating mice in a 2-3°C incubator for 8 h with food and water decreased the rectal core temperatures only from 36.5°C to 34.0°C (Cullen and Sarge, 1997). As mammals are multicellular organisms and have developed means to ward off the cold circumstances, their cold-response as a whole animal should naturally be different from bacteria and plants. When cultured as a single cell, however, some responses may be common to other single-cell organisms. In this review, I will focus on the studies using cultured mammalian cells and summarise what is currently known about the effects of cold exposure on cellular functions, especially on expression of cold-inducible genes.

Physiological Responses to Cold

For decades, cellular biologists have known that mammalian cells cultured at lower temperatures grow slower than those at 37°C. For example, mouse leukemic cells (L5178Y) grow exponentially after an initial lag phase (Watanabe and Okada, 1967). After 50-100 h, the culture reaches a stationary phase, in which there is no further increase in cell number. As shown in Figure 1, with decreasing temperature from 37 to 28°C, cell proliferation gradually decreases. The mitotic index of the exponential growth phase is constant for each culture and the value becomes smaller as the temperature falls. Similarly, in rat hepatoma Reuber H35 cells, approximately 95% remain undivided at 25°C, and after 48 h hypothermic death is 50-65% (van Rijn *et al.*, 1985). Until recently, many have thought that this decrease in growth rate is entirely due to the cold-induced depression of metabolism.

What phase of cell cycle is affected by the cold temperature? The G_1 phase seems to be the most severely affected of the four phases of cell cycle, although other phases are also affected to varying degrees. In rat H35 cells cultured at 25°C, G_1-phase cells do not appear to progress, while S-phase cells slowly proceed and are captured in a G_2 block (van Rijn *et al.*, 1985). In human amnion cells, Sisken et al. (1965) found that G_1 and M phases are most sensitive at temperatures below optimum, but all phases of the cell cycle respond quickly to changes in temperature. Chinese hamster fibroblasts incubated at 6 or 15°C are arrested in mitosis, while those incubated at 25°C accumulate in G_1 (Shapiro and Lubennikova, 1968). In other cell lines somewhat different patterns of temperature-dependent cell

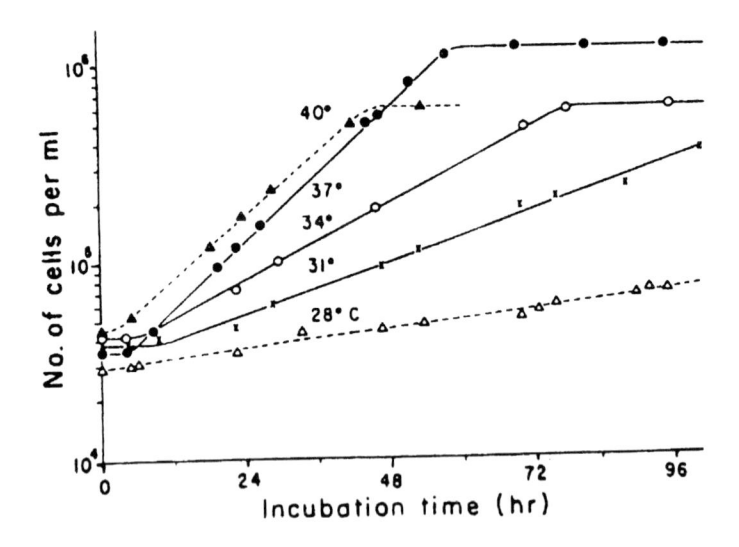

Figure 1. Growth Curves at Various Temperatures. The growth curves of mouse leukemia cell line, L5178Y, cultered at various temperatures. Note gradual decrease in the growth rate as the temperature decreases from 37 to 28°C. (Watanabe and Okada, 1967).

cycle progression has been reported (Rao and Engelberg, 1965; Watanabe and Okada, 1967), which may be due to the cells and temperatures used, time of analysis after temperature shift, and methods of analysis. When we analyzed the mouse BALB/3T3 fibroblasts cultured at 32°C by flowcytometry, the G1 phase was the most prolonged but other phases were affected as well (Nishiyama et al., 1997b).

Obviously, protein synthetic activity is necessary for animal cells to grow. When human HeLa cells are placed at 4°C, there is a gradual decline in their ability to synthesize protein (Burdon, 1987). If the cells are first subject to hyperthermic protocols that will induce HSP synthesis and tolerance to heat, no protective effect, but rather a more deleterious effect of cold (4°C) exposure on protein synthesis occurs, although one recent study using human IMR-90 fibroblasts demonstrates induction of tolerance to cold by previous heat shock (Russotti et al., 1996). Return of the HeLa cells to 37°C after a short exposure at 4°C permits recovery to normal levels of protein synthesis, but again these cells show no increased cold resistance. Lipid peroxidation is unlikely to be a direct cause of loss of protein synthetic function after hypothermic exposure (Burdon, 1987). Loss of cytoskeletal integrity is a possibility. Cytochalasin causes the release of mRNA from the cytoskeleton framework and inhibits protein synthesis (Ornelles et al., 1986), although its effects on translation may differ from those due to cold shock (Stapulionis et al., 1997). At 4°C, cultured cells tend to become more spherical. This change is paralleled by the sequential dismantling of the internal structure of the cell. Initially, the microtubules disassemble followed by the microfilaments (Weisenberg, 1972; Porter and Tucker, 1981). A number of translation components colocalize with cytoskeletal structures, and the loss and recovery of protein synthetic activity in Chinese hamster ovary (CHO) cells coincide closely with the F-actin levels. Disruption of actin filaments, but not microtubules, leads to a major reduction in protein synthesis, suggesting that the actin filaments are directly required for optimal protein synthesis (Stapulionis et al., 1997). The loss of activity can be reversed by a short recovery period under conditions that allow energy metabolism to occur; transcription and translation during the recovery periods are not necessary. Since the mammalian protein synthetic machinery is highly organized in vivo (Negrutskii et al., 1994), cold stress probably alters the supramolecular organization of this system, especially a portion of the microfilament network, and affects protein synthesis. The sequence of events leading from cold shock to an effect on protein synthesis is unknown, but one proposed scenario is as follows (Stapulionis et al., 1997); cold shock induces a transient permiabilization of the cells, which leads to an efflux of K^+ ions, an import of Na^+ and H^+ ions, and a slight reduction in cellular pH. These changes modulate the interaction of EF-1a with the actin cytoskeleton and affect translation.

The survival of Chinese hamster lung cells (V79), as measured by colony-forming ability, decreases below 37°C, but varies inversely with the temperature in the 10-25°C range and the macromolecular synthesis rate (Nelson et al., 1971). At lower temperatures, freezing and prefreezing damages occur, and MUTU-Burkitt lymphoma (BL) cells cultured on ice for 24 hr exhibit the cytological characteristics of necrosis: plasma-membrane rupture, disruption of cytoplasmic organelles, and absence of condensed chromatin (Gregory and Milner, 1994). By contrast, shorter periods in the cold induce apoptosis selectively. Group-I BL-derived cell lines, which retain in vitro the proliferative and apoptotic capacities of the parental cells, selectively enters apoptosis when returned to 37°C after a brief period, as little as 20 to 30 min, at 1°C or 4 hr at 25°C (Gregory and Milner, 1994). The induction of apoptosis as determined by morphological characteristics and DNA fragmentation is detectable within the first 1 to 2 hours of recovery at 37°C. The Bcl-2-dependent and -independent survival pathways are shown to provide protection from the cold-induced apoptosis, provided that these are active before exposure to low temperture. Since high levels of apoptosis are also inducible in group-1 BL cells by inhibitors of RNA and protein synthesis, the continued synthesis of one or more critical survival proteins of short half-life appears to be necessary to circumvent their apoptotic program. The cold-shock probably disturbs production of these proteins, leading to apoptosis. In addition, cold may stimulate synthesis of apotosis-inducing factor(s) as well (Grand et al., 1995). Cold-induced apoptosis is not observed in all BL cells; BL cells expressing Bcl-2 or tissue inhibitor of metalloproteinases (TIMP)-1 are resistant to the cold-induced apoptosis (Gregory and Milner, 1994; Guedez et al., 1998). Cold-induced apoptosis may also be dependent upon tissue of origin, stage of differentiation and/or cellular millieu, and has been described in other cells such as murine P815 mastocytoma (Liepins and Younghusband, 1985), BW5147 thymoma (Kruman et al., 1992), Chinese hamster V79 fibroblasts (Soloff et al., 1987) and human McCoy's synovial cells (Perotti et al., 1990). In these studies, effects of RNA/protein-synthesis inhibitors are not consistent, and apoptotic cell death appears to be dependent upon intracellular Ca^{2+} levels (Perotti et al., 1990), cell-cycle phase (Soloff et al., 1987; Perotti et al., 1990) and cytoskeletal stability (Liepins and Younghusband, 1985; Kruman et al., 1992). Further studies are required to establish how and to what extent cold shock treatment disrupts these processes and induces apoptosis.

At morphological, biochemical or molecular levels, several changes besides cytoskeletal changes have been observed in mammalian cells after cold exposure. Such changes are most probably due to the effects of low temperature on the physical properties of molecules and on rate processes. For example, phase transitions in the lipid bilayer occur temperature-dependently, and correlate with water permeability (Rule et al., 1980), glycosylation (Setlow et al., 1979), and adhesiveness of the membrane (Deman and Bruyneel, 1977). Changes in unsaturated fatty acid content of the cell membranes, which allow for the alteration of membrane fluidity in response to temperature shift, are induced and possibly related to the reduced proliferation of cells at 15°C (Shodell, 1975). Various rate processes such as diffusion, transport and enzyme activities will be affected. Most notably, protein unfolding, dissociation and inactivation will be induced by changes in hydrophobic interactions, ionization constants of charged groups on amino acid side chains and others (King and Weber, 1986)

Other cold-induced changes reported include the translocation of ß crystallin from the nuclear region into the cytoplasm (Coop et al., 1998), and tyrosin phosphorylation of p38 MAP kinase which is known to be activated by proinflammatory cytokines and environmental stresses (Gon et al., 1998). Clinically employed hypothermia is associated with bleeding diathesis, but hypothermia (33°C) does not adversely affect platelet functions, and rather increases intrinsic platelet reactivity by enhancing the exposure of activated GPIIb-IIIa receptors (Faraday and Rosenfeld, 1998). It probably reduces the availability of platelet activators. Obviously, more molecular changes are to be identified to explain the various biological changes induced by low temperatures.

Modification of Stress Responses

Effects on Recovery from Stress
Does hypothermia affect recovery of cells from stress? Post-treatment incubation of CHO cells at 0-4, 20, or 40°C has a differential influence on the expression of sublethal and potentially lethal damages due to hyperthermia or X-rays (Henle and Leeper, 1979).

Phillips and Tolmach (1966) studied the effect of low temperature (29°C) on the recovery of HeLa cells from X-rays, and observed a slight but reproducible decrease in survival by the hypothermic treatment. When confluent Chinese hamster cells are irradiated, the temperature dependence of the recovery is striking: after 6 h of recovery at 37, 25, and 4°C, the numbers of surviving cells increase 16-, 10-, and 2-folds, respectively, compared with the numbers immediately after irradiation (Evans et al., 1974). In other cell lines conflicting results have been reported. For example, the radiosensitive variant of the L5178 mouse leukemia cells irradiated in G1 phase of the cell cycle shows marked increase in survival as the postirradiation temperature is decreased through the range of 37 to 31°C (Ueno et al., 1979). The phase of cell cycle at irradiation is related to this hypothermic effect, although the onset of DNA degradation is delayed by hypothermia in cells at any phases.

Using an *in vitro* brain slice technique, cerebroprotective effect of hypothermia against repeated hypoxia has been demonstrated (Wassmann et al., 1998). Once hypoxia has occurred under normothemic conditions, no protective effect of hypothermia is observed, consistent with a notion that hypothermia preserves ATP stores. Hypothermia is also supposed to suppress rates of the generation of free radicals and other bioactive substances. However, as pointed out by Hochachka (1986), hypothermia-sensitive mammalian cells cannot maintain the regulated metabolism and the ion gradients across the membrane at low tempertures. Thus, complications will arise during a prolonged hypothermia treatment in the clinic.

Effects on Subsequent Stress Response

Cells grown at high temperature sometimes show stress response different from those grown at 37°C. Compared to the numerous studies on hyperthermia, hypothermia in combination with other treatments have received little attention. Chinese hamster HA-1 cells grown at 32°C are more sensitive to hyperthermia, but the response to X-rays is not affected (Li and Hahan, 1980). In other studies incubation at 4°C immediately prior to X-rays has no significant effect (Henle and Leeper, 1979; Holahan et al., 1982) or sensitization (Johanson et al., 1983). When the effects of prolonged incubation at suboptimal temperatures are examined extensively, significant effects become apparent (van Rijn et al., 1985). Rat H35 hepatoma cells are incubated at 8.5 °C or between 25 and 37°C for 24 h prior to hyperthermia or irradiation. Hypothermia causes sensitization to both treatments. Maximum sensitizaion is observed between 25 and 30°C and no sensitization is found at 8.5°C. These enhanced sensitivities disappears in approximately 6 h after return to 37°C. The mechanism for the observed hypothermic radio- and thermo-sensitization, a rather slow process, is not understood.

Brief exposure of cells to heat makes them heat tolerant. Thermotolerance can be induced by stresses other than heat, including cytotoxic drugs and ischemia, suggesting that thermotolerance represents a generalized response of cells to stress. Does it make cells cold tolerant? Heat shock at 42.5°C for 5 h improves survival of human IMR-90 fibroblasts to subsequent 4°C cold exposure, and the tolerance correlates with the induction of HSP27 (Russotti et al., 1996). When HeLa cells are first subject to short exposure at 4 or 45°C, however, no increased cold resistance in protein synthesis appears (Burdon, 1987). Induction of cold tolerance by hypothermia seems difficult, but not impossible. Pretreatment by 25-37°C cycling for more than

2 days makes V79 Chinese hamster cells more resistant to cold (5°C) as well as heat (43°C) (Glofcheski et al., 1993). If the pre-exposure is at 15 or 10°C, the resistance to hyperthermia is significantly reduced.

Because hypothermia either singly or accompanied by cardioplegia is regularly employed in myocardial protection during heart surgery, Ning, et al. (1998) examined the effects of hypothermia (31°C) in a perfused organ, rabbit hearts. Hypothermia preserves myocardial function and ATP stores during subsequent ischemia and reperfusion. Signaling for mitochondrial biogenesis is also preserved and *HSP70-1* mRNA is induced. To what extent these effects are due to hypothermia *per se* remains to be determined.

The outermost layer of the body, the skin is easily exposed to cold and sunlight. When ear skin of mice is exposed to cold stress at 0°C for 20 min or 5°C for 24 h and then exposed to UVB radiation, sunburn cell production is less than controls without cold exposure (Ota, et al., 1996). Sunburn cells are recognized as dead epidermal cells or apoptotic cells. The protective effects are also observed *in vitro*. Rat HT-1213 keratinocytes are exposed to 0°C for 1h and cultured further at 37°C. After 6-h incubation, they become resistant to cytotoxic effect of UVB. Since the level of metallothionein is increased after the cold stress and recovery, its radical-scavenging activity might contribute to photoprotection against UVB-induced oxidative damage (Ota, et al., 1996).

Gene Expression Induced by Cold Stress

Effects of Severe Cold Stress (Below 5°C)

If cold shock induces apoptosis in BL cells and protein synthesis is necessary for apoptosis, what gene(s) is induced? The protein named as apoptosis specific protein (ASP) is induced in BL cells undergoing apoptosis by a variety of stimuli including cold shock (4°C), while the level is low in cells dying by necrosis after prolonged exposure to cold (Grand et al., 1995). ASP is a cytoplasmic protein, and colocalizes with non-muscle actin. The recent isolation of *ASP* cDNA has revealed that it shows high homology to the *Saccharomyces cervisiae APG5* gene which is essential for autophagy, and that its mRNA levels are comparable in viable and apoptotic cells (Hammond et al., 1998). This posttranscriptional induction of ASP is probably a response to apoptosis rather than to the cold shock *per se*.

HSP is induced by various stresses in addition to heat. Effects of cold on induction of HSPs have been analyzed in many cell types including human skin biopsies, SCC12F squamous cell carcinomas (Holland et al., 1993), neutrophils (Cox et al., 1993), IMR90 fibroblasts, HeLa cells (Liu et al., 1994), and rat primary cardiomyocytes (Laios et al., 1997). Human skin biopsies are immediately exposed to 4, 15, 20, and 37°C for 1 h and then allowed to incorporate ^{35}S-methionine at 37°C for up to 3 h (Holland et al., 1993). At 15 or 4°C, increased synthesis of HSP72 and HSP90 is observed. In SCC12F cells, HSP72, but not HSP90, is induced after exposure to 4°C for 1 h. Liu et al. (1994) found that IMR-90 and HeLa cells preincubated at 4°C for 2-4 h followed by recovery at 37°C for 5 h synthesize and accumulate HSP98, 89 and 72, and that the degree of the induction is directly related to the time that the cells spend at 4°C. This induction is transcriptional because increase in the level of *HSP70* mRNA and activation of heat shock factor (HSF) are also observed. Without recovery at 37°C, however, no induction is observed. Laios et al. (1997) have studied the effect of hypothermia (4°C, 1 h) on the induction of HSP in primary cultures of rat cardiomyocytes. After recovery at 37°C for 2 h, induction of HSP70 was observed. The levels were maximal 4-6 h after recovery and began to decrease after 6 h. They also examined the effect at 4, 10, 15, 20 or 25°C with 4-h recovery at 37°C, and observed the induction of HSP70 under all conditions. Although the increase in protein levels correlates with induction of mRNA for HSP70, only mRNA is induced for HSP25, and no induction is observed for HSP90. These results suggest that cold induction of HSP is differentially regulated at the protein and mRNA levels and that individual HSP is regulated differently from others even in the same cell type. These studies are consistent with the notion that the sensing mechanism of the heat shock response detects a relative, as opposed to absolute, temperature change. In other studies, however, the effects of cold are highly variable from induction to suppression according to many factors possibly including cell types, cell cycle status, and temperature and duration of preincubation used (Hatayama et al., 1992).

Cold stress causes various inflammatory processes. Although cold exposure induces phosphorylation of p38 MAP kinase in NCI-H_{292} cells, IL-8 is induced only after rewarming to 37°C for 6 h (Gon et al., 1998). Since IL-8 is known to cause airway inflammation, the p38 MAP kinase-dependent pathway may be causally related to the exercise-induced bronchoconstriction and/or the cold preservation and rewarming-induced injury of transplant organs.

WAF1^{p21}/CIP1/sdi1 is a major mediator for p53-dependent G_1 arrest (El-Deiry et al., 1993). A-172 human glioblastoma cells are incubated at 4, 15, or 20°C for 1h followed by recovery at 37°C for 10 h (Ohnishi et al., 1998). Accumulation of p53 and WAF1 is induced, which is inversely correlated with the treatment temperature. Since *WAF1* mRNA, but not *p53* mRNA is increased by cold stress, p53 accumulation is due to post-transcriptional events as observed in UV or radiation-induced stress response.

In all cases described above, it is not clear whether or not these genes are induced as a result of the metabolic stress caused by cold shock or the temperature up-shift from 4 to 37°C, namely a heat shock response. Given that cold can denature protein and denatured protein can induce the heat shock response, it would seem reasonable to assume that induction of the genes including HSPs is a cold shock response. The observation that the onset, magnitude, and duration of the induced response is directly proportional to the severity of cold shock is consistent with this hypothesis (Liu et al., 1994). However, although the rapid temperature change to the normal physiologic one is unlikely to promote the denaturation of proteins, it remains as a possibility. Furthermore, mechanisms other than protein denaturation might be activated by the temperature up-shift to induce the genes observed. For example, the sudden resumption of active cell

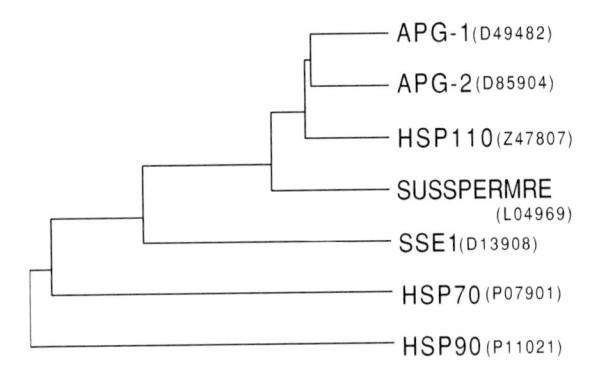

```
            ┌── APG-1 (D49482)
          ┌─┤
          │ └── APG-2 (D85904)
        ┌─┤
        │ └──── HSP110 (Z47807)
      ┌─┤
      │ └────── SUSSPERMRE
      │               (L04969)
    ──┤──────── SSE1 (D13908)
      │
      │ ┌────── HSP70 (P07901)
      └─┤
        └────── HSP90 (P11021)
```

Figure 2. Phylogenetic Relationships. Phylogenetic relationship of mouse APG-1 to various HSP. In mice and humans, the HSP110 family is consisted of APG-1, APG-2 and HSP110/105. SUSSPERMRE: sea urchin egg receptor for sperm. Accession number of each of the *HSP* cDNAs is given in parenthesis.

metabolism and increased production of oxygen-free radicals may contribute to the induction. Especially since no gene has been reported to be induced at the low temperature (below 5°C) without recovery at higher temperatures, this question is difficult to be answered

Effects of Mild Cold Stress (32°C)
At 4°C, active cell metabolism such as ATP production and macromolecular synthesis is not observed. With mild hypothermia, however, cells can survive and proliferate. Chinese hamster HA-1 cells that have been maintained at 37°C for several years can adapt themselves to grow at temperatures ranging from 32 to 41°C (Li and Hahn, 1980). This growth adaptation is accompanied by various changes including the cholesterol/phospholipid ratios of plasma membranes. In mouse FM3A cells, the inducibility of HSP70 by the 42°C-heat shock is lost, while that of HSP105 by the 39°C-heat shock is maintained after cultivation at 33°C for more than a month (Hatayama *et al.*, 1992).

In many mammals including humans, the testes descend into a scrotum during fetal or early post-natal life (Setchel, 1982). The testis finds itself in a cooler environment, usually between 4 and 7°C cooler, and then germ cells become susceptible to damage if testicular temperature is raised to that of the body cavity. In experimental animals, surgical induction of cryptorchidism or exposure of the testis to heat causes apoptosis of germ cells, especially primary spermatocytes, within 2-4 days, leading to infertility (Chowdhury and Steinberger, 1970; Nishiyama *et al.*, 1998a). Various exogenous thermal factors including those observed in paraplegic patients in wheelchairs and welders are proposed to be risk factors for human male infertility. Cryptorchidism and varicocele of the spermatic veins are associated with male infertility, and their pathogenesis are attributed to thermal factors. Although male germ cell-specific alteration in temperature set point of HSF1 activation is suggested (Sarge, 1995), the molecular mechanisms of the thermal effect on spermatogenesis are just beginning to be explored.

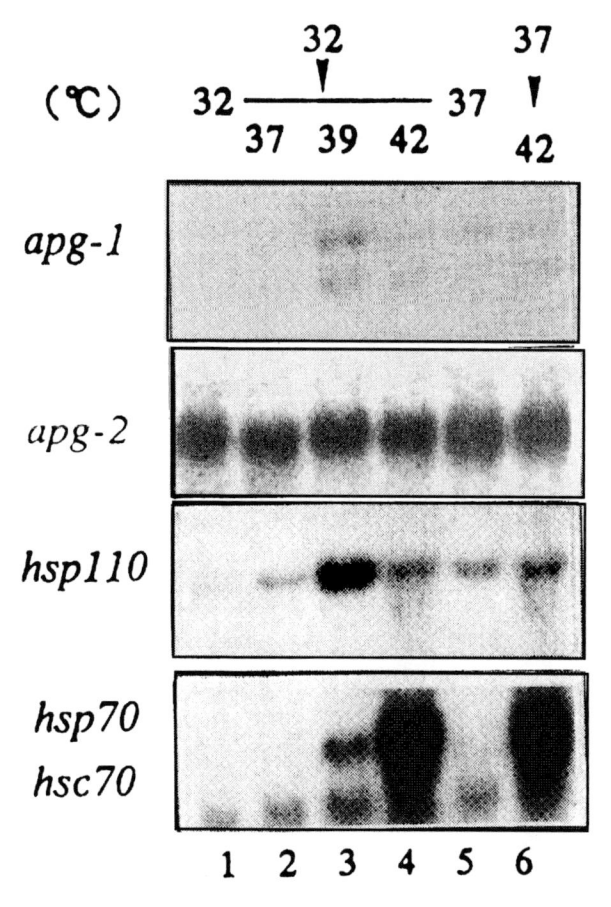

Figure 3. Heat Induction of HSP Transcripts. Heat induction of the various HSP transcripts in TAMA26 Sertoli cells. TAMA26 cells were heat-shocked at the indicated temperatures for 2 h after growth for 20 h at 32 or 37°C. The same filter was hybridized with the indicated cDNA probes consecutively. (Kaneko et al., 1997a, 1997c).

In an attempt to identify genes involved in spermatogenesis, we subtracted the testis cDNAs of the prepubertal mice from those of adult mice (Kaneko et al., 1997c). One novel cDNA encoding a protein with an ATP-binding motif and peptide-binding motif was isolated and named as APG (Ath and Peptide-binding protein in Germ cells)-1. After publishing APG-1 in the database, the same gene was identified as OSP94 (Kojima et al., 1996). It has been long recognized that the major HSPs of mammalian cells are observed at 28, 70, 90, and 110 kDa and other HSP families, e.g. HSP60 and HSP40, have been subsequently identified. Recently, the cloning of *HSP110* cDNA from hamster, mouse, yeast, arabidopsis, and a variety of other species has been described (Easton and Subjeck, 1997). The HSP110 family is a significantly enlarged and diverged relative of the HSP70 family with unique sequence components (Oh et al., 1999), and includes as members a sea urchin egg receptor for sperm and yeast SSE1 (Figure 2). Together with APG-2 (Kaneko et al., 1997a) and HSP110/105 (Yasuda et *al.,* 1995), APG-1 constitutes the HSP110 family in mice and humans. When mouse NIH/3T3 fibroblasts or TAMA26 Sertoli cells maintained at 37°C are exposed to 42°C for 2 h, no or only a slight induction of *APG-1*, *APG-2* and *HSP110* mRNAs is observed, while *HSP70* mRNAs are markedly induced under the same conditions (Figure 3, Kaneko *et al.,* 1997a, 1997c). The temperature of induction of the heat shock response is related to the physiological temperature range to which the cells and organisms are adapted (Lindquist and Craig, 1988). Since *APG-1* mRNA is constitutively expressed mainly in the testis (Kaneko *et al.,* 1997b), TAMA26 cells are first exposed to the normal testicular temperature, 32°C, for 20 h and then incubated at 37, 39, or 42°C for 2 h. As shown in Figure 3, *APG-1* mRNAs are induced by a temperature shift from 32 to 39°C (low temperature heat shock), but not by a shift from 32 to 42°C or from 32 to 37°C. The heat response pattern of *HSP110* expression is similar to that of *APG-1*. Although induction of *HSP70* transcripts is observed in 2 h by a shift from 32 to 39°C, the induction is more apparent by a shift from 32 to 42°C or from 37 to 42°C. No induction of *APG-2* is observed under any conditions. Essentially similar differential response patterns are observed among these genes in other mouse cell lines and human cell lines. These findings suggest that the mechanisms regulating the levels of *APG-1* and *HSP110* are different from those of *HSP70* or *APG-2* (Kaneko *et al.,* 1997a, 1997c; Xue *et al.,* 1998). More importantly, they definitely demonstrate that exposure to mild hypothermia (32°C) can modify the subsequent response of the cells to mild fever-range hyperthermia (39°C) and induce distinct gene expressions. Whether the changes in the membrane's physical state (Vigh *et al.,* 1998) are related to these observations remains to be determined.

The induction of genes described above is phenomenologically distinct from the regulation of cold-shock genes in bacteria and plants. The induction of cold shock proteins is a response to the lowering of the culture temperature itself and occurs at the low temperature. If mammalian testicular germ cells actively divide and differentiate at 32°C better than at 37°C, it is highly probable that there are genes specifically expressed at 32°C in them.

CIRP, a Novel Cold Shock Protein in Mammals

Proteins with RNA-binding motifs are known to play important physiological roles in humans and other organisms. For example, Tra2, Elab, and Rb97D are essential in *Drosophila* for the sex determination, development and maintenance of neurons and spermatogenesis, respectively (Amrein *et al.,* 1988; Robinow *et al.,* 1988; Karsch-Mizrachi and Haynes, 1993). The Y-chromosome genes *YRRM/RBM1* (Ma *et al.,* 1993) and *DAZ* (Reijo *et al.,* 1995) are candidate genes for the azoospermia factor that controls human spermatogenesis. To identify genes involved in spermatogenesis we screened for RNA-binding proteins (RBPs) expressed in the mouse testis with a PCR-based cloning method, and isolated a cDNA encoding a novel RBP designated as CIRP (cold-inducible RNA-binding protein) (Nishiyama *et al.,* 1997b).

Structure of CIRP

The mouse *CIRP* cDNA encodes protein of 172 amino acids which displays two main features: the presence of an amino-terminal consensus sequence RNA-binding domain (CS-RBD), and a carboxyl-terminal glycine-rich domain. CS-RBD, also referred to as ribonucleoprotein (RNP) motif, RNP consensus sequence, or RNA recognition motif, is one of the major RNA-binding motifs and is the most widely found and best characterized (Burd and Dreyfuss, 1994). It is composed of ~90 amino acids, including two highly conserved sequences, an octamer designated RNP1 and a hexamer designated RNP2, and a number of other, mostly hydrophobic conserved amino acids interspersed throughout the motif. The CS-RBD of CIRP contains a consensus sequence of RNP1 and RNP2, and a number of highly conserved segments which could be perfectly aligned with those of other CS-RBDs from proteins of divergent organisms (Figure 4). Most RNPs contain multiple CS-RBDs, but CIRP contains only one, and its carboxyl-terminus is rich in glycine, serine, arginine, and tyrosine (38.8, 16.4, 19.4, and 10.4%, respectively). The gycine-rich domain is supposed to enhance RNA-binding via protein/protein and /or protein/RNA interactions (Dreyfuss *et al.*, 1993). In addition, dimethylation or phosphorylation of these residues may affect the function of CIRP. The carboxyl-terminal auxiliary domain contains several repeats of the RGG sequence, the so-called RGG box (Burd and Dreyfuss, 1994). The RGG box was initially identified as an RBD in hnRNP U, and usually occurs in proteins that also contain other types of RBDs. It can increase RNA affinity of other RBDs nonspecifically in addition to the sequence-specific RNA binding activity.

The overall amino acid sequence of CIRP is identical to rat CIRP (Xue *et al.*, 1999), and shows the highest similarity by FASTA3 search of the database to human CIRP (Nishiyama *et al.*, 1997a; Sheikh *et al.*, 1997), followed by frog CIRP (XCIRP, Uochi and Asashima, 1998), axolotl RBP (Bhatia *et al.*, 1998), mouse RBM3 (Danno *et al.*, 2000), human RBM3/RNABP (Derry *et al.*, 1995), and plant glycine-rich RNA-binding proteins (GRPs)(van Nocker and Vierstra, 1993; Dunn *et al.*, 1996; Bergeron *et al.*, 1993; Carpenter *et al.*, 1994) (Figure 4). In particular, CIRP is 50-80% identical in the CS-RBD to these proteins and also contains the sequences DRET and MNGKXXDG which are highly conserved in RBPs with only one CSRBD. Phylogenetic analysis of these proteins suggests two distinct groups: animal and plant, and the animal RBP clusters display two subclasses: mouse CIRP/human CIRP/human RBM3/axolotl RBP and human hnRNPG/mouse hnRNP (Bhatia *et al.*, 1998).

Induction by Cold Stress

Two *CIRP* transcripts, a major transcript of 1.3 kb and a faint transcript of 3.0 kb in size, are constitutively expressed in the testis, brain, lung and heart, and at lower levels in most tissues of adult mice (Nishiyama *et al.*, 1998a). Since expression of some plant GRPs are inducible by cold stress (Bergeron *et al.*, 1993; Carpenter *et al.*, 1994; Heintzen *et al.*, 1994; Dunn *et al.*, 1995) and *CIRP* is expressed in the testis where it is cooler than the

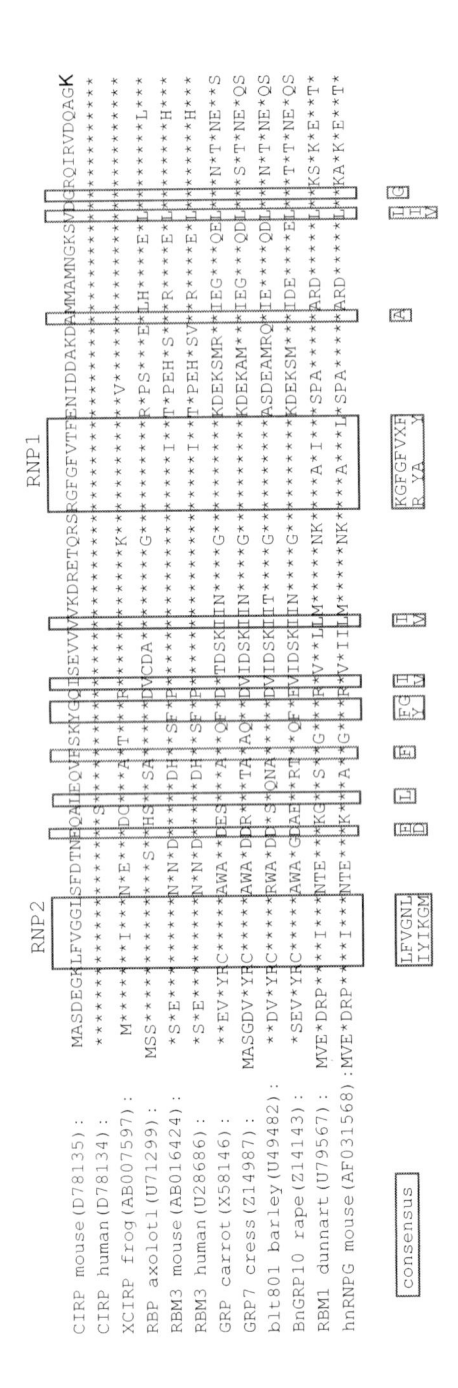

Figure 4. Protein Alignments. Alignment of the CS-RBD (RNP domain) in mouse CIRP with its related proteins. The proteins are placed in the decending order of overall amino acid sequence similarity to mouse CIRP. Consensus sequences found in CS-RBDs (Burd and Dreyfuss, 1994) are boxed. Accession number of each of the cDNAs is given in parenthesis.

(a) 15 25 32 37 39 42 (°C)

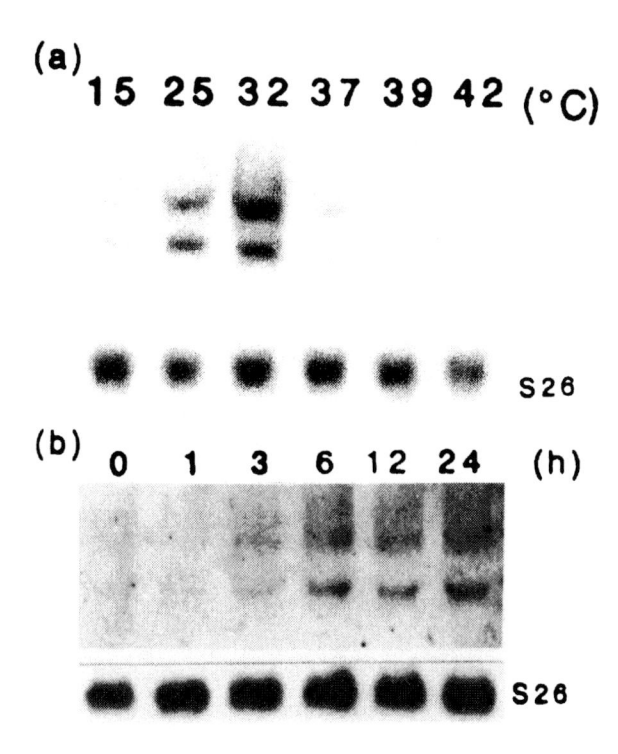

S26

(b) 0 1 3 6 12 24 (h)

S26

Figure 5. Cold Stress Induction of mRNA. Induction of *CIRP* mRNA by cold stress. (a) Temperature-dependence. Northern blot analysis of total RNAs from BALB/3T3 fibroblasts harvested 24 h after the indicated temperature shift from 37°C. (Nishiyama et al., 1997b) (b) Kinetics. Northern blot analysis of total RNAs from BALB/3T3 fibroblasts harvested at indicated times after the 37 to 32°C temperature shift. As a control for the amount of RNA loaded, the filter was rehybridized with a mouse S26 ribosomal protein cDNA probe.

body cavity, BALB/3T3 mouse fibroblasts maintained at 37°C was exposed to cold stress of various temperatures (Nishiyama et al., 1997b). At 24 h after transfer to 32°C, strong expression of *CIRP* mRNA and protein is induced, while incubation at 39 or 42°C decreases *CIRP* expression (Figure 5a). Similar effects of mild cold and heat is observed in all examined cell lines of diverse origins including mouse TAMA26 Sertoli cells, BMA1 bone marrow stromal cells, and human HepG2 hepatoma cells (Nishiyama et al., 1997a, 1997b). Thus, CIRP is the first cold shock protein identified in mammalian cells. It should be noted that severe cold stress (15°C and below) does not cause induction of *CIRP* (Figure 5a), which explains, at least partly, the absence of mammalian cold shock proteins reported in the literature. In BALB/3T3 cells, the induction of *CIRP* mRNA and protein becomes evident 6 h after a temperature down-shift from 37 to 32°C (Figure 5b).

Multiple Pathways to CIRP Induction

Cold shock may inflict cellular stress by affecting the structure and function of cytoplasmic enzymes, cytoskeletal proteins and membrane proteins. Specifically, cold shock may induce the unfolding, dissociation, and

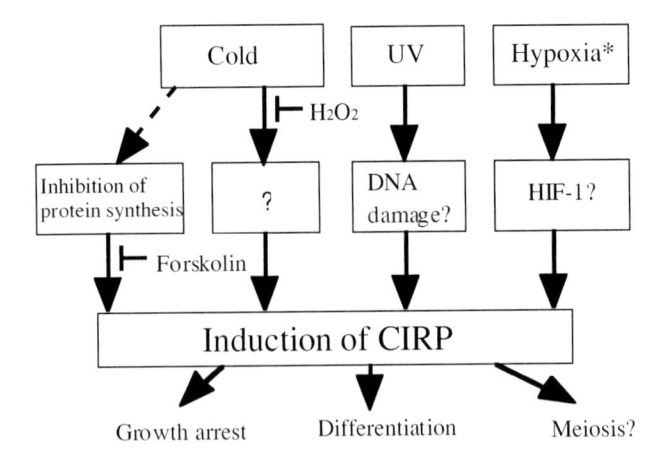

Figure 6. Cold Stress Induction of CIRP. Induction of CIRP by cold and other stresses. At least two different pathways exist; one cold-inducible and the other protein synthesis inhibitor-inducible, as demonstrated by the differential inhibitory effects of H_2O_2 and forskolin. In addition, UV irradiation and possibly hypoxia induce CIRP mRNA expression. *Demonstrated by whole animal exposure.

inactivation of cellular proteins (King and Weber, 1986; Pain, 1987; Watson and Morris, 1987). Since denatured proteins are known to induce the HSPs, similar mechanisms may be present for induction of cold shock proteins. Cold-induced changes in RNA secondary structure or molecular order of membranes may serve as a temperature sensor in the cell. Conformational changes of transcription factors may also be induced. At the present moment, however, we have little information concerning the nature of the sensing and signal transduction mechanisms of the cold shock response in mammalian cells. Therefore, we will consider specifically the induction mechanism of CIRP.

Hypothermia increases levels of CIRP mRNA and protein in cultured cells. Hypothermia delays degradation of mRNAs in general, but the change in the half-life of CIRP mRNA is not so different from those of other mRNAs (H. Nishiyama and J. Fujita, unpublished). Thus, the CIRP gene seems to be preferentially transcribed under the hypothermic conditions. Indeed, we have isolated the mouse CIRP gene containing the "cold responsive element", which confers on the reporter gene the cold (32°C) inducibility (H. Higashitsuji and J. Fujita, unpublished). Charcterization of the factor(s) that binds to this element will greatly facilitate our understanding of the cold shock responses in mammalian cells. In addition, a mechanism of translational control may be operative which both specifically promotes the translation of CIRP and other cold-shock protein mRNAs and suppresses the tranlation of other mRNAs. If so, there should be a trasnslational mechanism that exclusively recognizes CIRP mRNAs and/or excludes other mRNAs, and some specific structural features that differentiate CIRP mRNAs from other mRNAs.

In prokaryotes, the cold shock response is induced when ribosomal function is inhibited either by cold-sensitive ribosomal mutations, or by antibiotics (Thieringer et al., 1998). Many data suggest that the physiological signal for the induction of the cold shock response in microorganisms is inhibition of initiation of translation caused by the abrupt shift to lower temperature. In mammalian cells, inhibition of translation by cycloheximide or puromycine induces CIRP as well as RBM3 mRNAs (Danno et al., 1997; H. Tokuchi and J. Fujita, unpublished). This observation, combined with the observation that shifting cells to lower temperatures causes inhibition of protein synthesis (Burdon, 1987) suggests that the state of the ribosome is the physiological sensor for the induction of the cold-shock response in mammalian cells as well. Induction of CIRP mRNA by protein synthesis inhibitors is inhibited by forskolin, but not by H_2O_2. By contrast, the induction by cold is inhibited by H_2O_2, but not by forskolin (H. Tokuchi and J. Fujita, unpublished). These results suggest that signals from cold and protein synthesis inhibitors are trasduced via two independent pathways for induction of CIRP mRNA (Figure 6). Human CIRP cDNA has been independently isolated as a human homolog of A18, one of the hamster genes rapidly induced (within 4 h) by UV irradiation in CHO cells (Sheikh et al., 1997). UV and the UV mimetic agent AAAF, but not MMS and H_2O_2, up-regulate A18 mRNA in human cells at 37°C. DNA damage caused by the former agents is predominantly repaired via nucleotide excision repair pathway, while that by the latter is by base excision repair pathway (Sadaie et al., 1990). Different genotoxic agents including UV are found to activate the c-Abl and JNK wathways to differing extents, but p53 responds to every agent tested, independently of c-Abl or JNK (Liu et al., 1996). p53 is now regarded as an integration point for stress signals. Since p53 is involved in both growth arrest and apoptotic cell death (Gottlieb and Oren, 1998) and severe cold is shown to activate p53 (Ohnishi et al., 1998), the UV-induced signal transduction pathway may overlap with that of the cold response. Recently, we have observed induction of CIRP in many tissues, especially lung, when mice are kept under hypoxic conditions (S. Masuda and J. Fujita, unpublished). Whether this is mediated via the binding site for the hypoxia inducible transcription factor (HIF)-1 present in the CIRP gene or other pathways is under investigation. The regulatory mechanisms of the diurnal change in the expression of CIRP (Nishiyama et al., 1998b) is also unknown.

Function of CIRP

When incubation temperature is decreased from 37 to 32°C, the growth of cultured cells is impaired. Indeed, doubling time of BALB/3T3 cells prolongs from 19 h (37°C) to 29 h (32°C) (Nishiyama et al., 1997b). In BALB/3T3 cells the presence of antisense oligodeoxynucleotide (ODN) to CIRP mRNA in the culture medium partially inhibits the induction of CIRP by temperature downshift (Figure 7). Concomitantly, the growth impairment at 32°C is partially alleviated, suggesting that the induction of CIRP is necessary for growth suppression by cold stress. Furthermore, overexpression of CIRP at 37°C induces prolongation of the G1 phase of cell cycle and reduces the growth rate. Although some factor(s) other than CIRP also contributes to the impaired growth at 32°C, CIRP has a mitosis-inhibitory activity and plays an

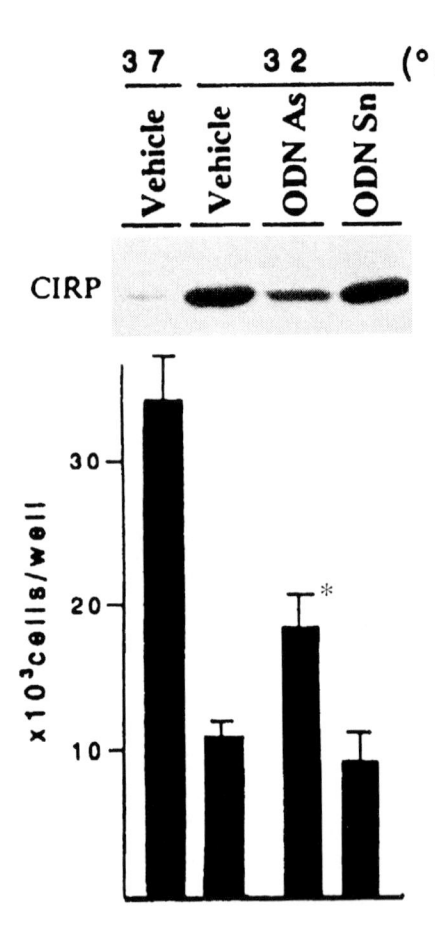

Figure 7. Antisense Suppression of CIRP. Effects of antisense (As) oligodeoxynucleotide (ODN) on the cold-induced suppression of cell growth. BALB/3T3 cells were incubated at the indicated temperatures in the presence of vehicle alone, As, or sense (Sn) ODN (0.5 μM). Expression of CIRP was analyzed by Western blotting after 12 h of culture (upper). Note suppression of the cold-induced CIRP expression in the presence of As ODN. Cell numbers were determined after 2 d of culture (lower). The results are expressed as the mean ± SEM. *Statistically different from controls ($P<0.02$). (Data from Nishiyama et al., 1997b, by copyright permission of the Rockefeller University Press.).

essential role in cold-induced growth suppression of mammalian cells. These findings demonstrate that the decreased growth rate of mammalian cells at lower temperature (Figure 1) is not entirely due to the arrested metabolism as has been thought, but involves an active process, induction of a cold shock protein.

In the mouse testis, CIRP mRNA and protein are constitutively expressed in the germ cells and the level varies depending on the stage of differentiation (Nishiyama et al., 1998a). Spermatogonia develop into primary spermatocytes after several mitotic divisions. In primary spermatocytes, mitotic cell division is suppressed, and DNA replication, homolog pairing, and recombination occur (Bellve, 1979). Subsequently, spermatocytes give rise to four haploid cells by two meiotic divisions. Mouse CIRP is strongly expressed in primary spermatocytes but not in spermatogonia, and mitotic proliferation of the GC-2spd(ts) germ cell is suppressed by CIRP. Taken together CIRP may be involved in suppression of the mitotic cell cycle after differentiation of spermatogonia to spermatocytes. The expression of CIRP in germ cells is heat sensitive; it decreases at the body cavity temperature (Nishiyama et al., 1998a). In experimental cryptorchidism, the decrease in the level of *CIRP* mRNA precedes the decrease of other gene expression and degeneration of germ cells. The earliest cellular damages are noticed in primary spermatocytes and early spermatids, in which CIRP is strongly expressed at scrotal temperature (32°C). In human testis with varicocele, CIRP expression is also decreased. As the timing of mitosis and meiosis in spermatogenesis is strictly controled, decreased expression of CIRP may adversely affect their coordinate regulation, and lead to disruption of spermatogenesis and apoptosis at 37°C.

In the brain but not in the liver and testis, the level of *CIRP* mRNA is diurnally regulated (Nishiyama *et al.*, 1998b). It increases during the daytime, reaching the highest level at 18:00, and then decreases to the lowest at 03:00. Interestingly ,this pattern of diurnal change is similar to that observed in some plant GRPs which are also cold inducible (Carpenter *et al.*, 1994; Heintzen *et al.*, 1994). In mammals, suprachiasmatic nuclei are considered to be the primary pacemaker for the circadian rhythms (Meijer and Rietveld, 1989). Immunohistochemical analysis shows that CIRP is expressed in the nucleus of neurons and that the level of CIRP is diurnally regulated in the suprachiasmatic nucleus and cerebral cortex. Thus, CIRP may play a role in biological rhythms.

From these observations CIRP seem to have at least two major functions. One is suppression of mitosis. The other is promotion or maintenance of differentiation. The finding that expression of CIRP coincides with cessation of mitosis and initiation of differentiation in neuronal cells in the developing mouse embryo (Sato and Fujita, unpublished) is consistent with this notion. Production of *CIRP*-gene knockout mice is in progress, and expected to clarify various functions of CIRP.

Proteins with CS-RBD are believed to be involved in posttranscriptional regulation of gene expression (Burd and Dreyfuss, 1994). In eukaryotic cells, mRNAs are produced in the nucleus from the primary transcripts of protein-coding genes (pre-mRNAs) by a series of processing reactions including mRNA splicing and poly-adenylation. In the cytoplasm,the translation and stability of mRNAs are also subject to regulation. Immunohistochemical analysis shows that the strong signals for CIRP is present in the nucleus of various cells including fibroblasts, neurons and germ cells in both mice and humans (Nishiyama *et al.*, 1997b, 1998a, b), suggesting that CIRP plays an important role(s) in RNA biogenesis in the nucleus. In addition, CIRP is present in cytoplasm but not in nucleus of round spermatids of mice (Nishiyama *et al.*, 1998a). The mechanisms regulating localization of CIRP and function of CIRP in haploid cells are of interest.

Binding experiments with ribohomopolymers and various RNAs have established that most of the CS-RBD-containing proteins have distinct RNA-binding characteristics (Burd and Dreyfuss, 1994). CIRP shows preferential binding to poly(U) *in vitro* (Nishiyama *et al.*, 1997b). Although structurally similar, axolotl RBP binds strongly with poly(A) and to a lesser degree with poly(U), but not with poly(G) or poly(C) (Bhatia *et al.*, 1998). blt801 has affinity for poly(G), poly(A) and poly(U) but not for poly(C)(Dunn *et al.*, 1996). RNP1 and RNP2 are not considered to distingush different RNA sequences.

Table 1. Cold-Induced Gene Expression in Cultured Mammalian Cells

Cold stress (Duration)	Recovery (Duration)	Cells (Species)	Induced genes	Reference
I. Severe cold stress				
4, 15°C (1 h)	37°C (2 h)	skin biopsies (human)	*HSP72, HSP90, HSP72*	Holland et al., 1993
		SCC12F (human)		
4°C (2-4 h)	37°C (4-6 h)*	IMR-90, HeLa (human)	*HSP70, HSP89, HSP98*	Liu et al., 1994
1°C (4 h)	37°C (4 h)	MUTU-BL (human)	Apoptosis specific protein (ASP), metallothionein	Grand et al., 1995
0°C (1 h)	37°C (6 h)	HT-1213 (rat)	*HSP70, HSP25*	Ota et al., 1996
4, 10, 15, 20, 25°C (1 h)	37°C (2 h)	primary cardiomyocytes (rat)		Laios et al., 1997
1°C (1 h)	37°C (6 h)*	NCI-H$_{292}$ (human)	*IL-8*	Gon et al., 1998
4°C (1 h)	37°C (10 h)	A-172 (human)	*WAF1*	Ohnishi et al., 1998
II. Mild cold stress				
32°C (20 h)	39°C (2 h)*	Balb/3T3, others (mouse)	*APG-1, HSP105, HSP70*	Kaneko, et al., 1997c
32°C (6 h)	none	Balb/3T3, others (mouse)	*CIRP*	Nishiyama, et al., 1997b
		T24, others (human)		Nishiyama, et al., 1997a
		PC12 (rat)		Xue et al., 1999
32°C (6 h)	none	T24, others (human)	*RBM3*	Danno, et al., 1997
		Balb/3T3, others (mouse)		Danno et al., 2000
32°C (6 h)	none	ECV304 (human)	*KIAA0058*	Sato and Fujita**

*no induction without incubation at indicated temperature, **unpublished observation.

Selectivity for specific RNA sequences are provided by variable regions of the CS-RBD and other domains. The physiological roles that RBPs with CS-RBDs play are unknown, but RNA bound to them is known to be relatively exposed and potentially accessible for interaction with other RNA sequences or RBPs (Burd and Dreyfuss, 1994). In fact, many RBPs with CS-RBD can destabilize the helix and/or promote annealing of complementary nucleic acids (Portman and Dreyfuss, 1994). Such activity could dramatically influence overall RNA structure and may be akin to chaperone activity. Thus, CIRP may function as an RNA chaperone, preventing secondary structure formation in RNA and/or modifying RNA/RNA annealing at low temperature as proposed for cold shock proteins in bacteria (Thieringer et al., 1998). Identification of target RNAs in vivo will help elucidate function of CIRP at the cellular and molecular levels. The possibility that CIRP has additional functions by interacting with other proteins via auxillary domain should also be investigated, especially when the mechanisms of suppression of G1 progression is yet to be clarified.

CIRP is expressed constitutively as well as after exposure to cold. In addition, CIRP can be induced by a wide range of stresses, further underscoring its biological importance. The continued study of CIRP may yield considerable insights relevant to fundamental biological phenomena, i.e. cell growth and differentiation.

More Cold-Inducible Genes

To find more cold-inducible genes (Table 1), two kinds of approaches are conceivable; candidate gene approach and a more comprehensive approach. RBM3 is structurely quite similar to CIRP with one amino-terminal CS-RBD and a carboxyl-terminal glycine-rich domain. Although the kinetics and degree of cold induction are slightly different from those of CIRP, RBM3 mRNA is increased after temperature downshift from 37 to 32°C in all human cell lines examined (Danno et al., 1997). The effects of protein synthesis inhibitors, cycloheximide and puromycine, are exactly the same as described for CIRP, namely dose-dependent induction of mRNA at 37°C. However, RBM3 is not involved in the cold-induced growth suppression (Danno et al., 2000), and tissue distribution of mRNA expression is different from CIRP. CIRP and RBM3 seem to be differentially regulated under physiological and stress conditions and serve distinct functions. Although hnRNP A1 is structurally related to CIRP and RBM3, it contains two CS-RBDs instead of one and is not induced by incubation at 32°C (Danno et al., 1997).

There is no formal definition of what constitutes a candidate gene for cold-inducible gene, and more genes remain to be analyzed. For example, the amino acid sequences of several eukaryotic transcription factors (Y-box proteins) are related to the bacterial cold domain (Burd and Dreyfuss, 1994). Members of the DEAD box subfamily of RNA helicases have been isolated from mammals (Luking et al., 1998). As known for prokaryotes, many eukaryotic organisms also have a homeoviscous response at the cellular level (Thieringer et al., 1998), and several rodent and human lipid desaturases have been isolated.

Technological advances have made it possible to compare gene expression profiles in different physiological/pathological situations, and the analysis of the kinetics of multiple gene expression after exposure to stress is now feasible. cDNA microarrays are increasingly utilized for this purpose. By this technique, we have compared the expression profiles of about 8000 genes in human cells cultured at 32°C and 37°C (T. Sato and J. Fujita, unpublished). One of the genes identified is KIAA0058, originally isolated by analysis of cDNA clones from a human immature myeloid cell line KG-1 (Nomura et al., 1994). Its function in cold stress response is currently under investigation. A rapid increase in the discovery of cold-inducible genes is expected to occur.

Conclusion

In hypoxia-tolerant animals, metabolic arrest by means of a reversed or negative Pasteur effect and maintaining membranes of low permeability are the most effective strategies for extending tolerance to hypoxia (Hochachka, 1986). The clinical application of this strategy is not easy, mainly because depression of metabolism through cold is the usual arrest mechanism used, and hypothermia in itself perturbs controlled cell function. In the long-term, hypothermia causes serious complications, such as respiratory blockade, heart failure, and infection (Wassmann et al., 1998). By clarifying the molecular mechanisms of cold response, a novel strategy for prevention and cure of ischemic damages without complications might be developed.

Not only in the clinic, but also under normal physiologic conditions mild hypothermia is widely experienced. The testis temperature is 30-34°C. The mean skin temperature is around 33°C, and becomes 27°C after immersion for 1 h in a 24°C-water bath (Marino and Booth, 1998). The temperatures in some organs that contact the external environment such as the respiratory and digestive tracts are lower than the body cavity temperature (H. Nishiyama and J. Fujita, unpublished). We know little about the physiological significance of these mild cold stresses.

I have shown in this review that at 32°C a cold-stress response is elicited and several genes are activated in mammalian cells. No doubt many more are yet to be discovered. The beneficial effects of hypothermia, e.g. in the treatment of brain damage, may not be entirely due to the depressed metabolism. Active metabolism continue in cells at 32°C, and cells may survive and respond to stresses with different strategies from those at 37°C. The study of the cellular and molecular biology of mammalian cells at 32°C is a new area which can be expected to have significant implications for medical sciences and possibly biotechnology.

Acknowledgements

I thank all the collaborators, especially Drs. Yoshiyuki Kaneko and Hiroyuki Nishiyama for helpful discussion, and Dr. R. John Mayer, University of Nottingham Medical School for a critical reading of the manuscript.

References

Amrein, H., Gorman, M., and Nothiger, R. 1988. The sex-determining gene tra-2 of Drosophila encodes a putative RNA binding protein. Cell. 55: 1025-1035.

Bellve, A. 1979. The molecular biology of mammalian spermatogenesis. In: Oxford Reviews of Reproductive Biology, vol 1. C.A. Finn, ed. Oxford Univ Press, London. p. 159-261.

Bergeron, D., Beauseigle, D., and Bellemare, G. 1993. Sequence and expression of a gene encoding a protein with RNA-binding and glycine-rich domains in Brassica napus. Biochim. Biophys. Acta. 1216: 123-125.

Berntman, L., Welsh, F.A., and Harp, JR. 1981. Cerebral protective effect of low-grade hypothermia. Anesthesiology. 55: 495-498.

Bhatia, R., Gaur, A., Lemanski, L.F., and Dube, D.K. 1998. Cloning and sequencing of the cDNA for an RNA-binding protein from the Mexican axolotl: binding affinity of the *in vitro* synthesized protein. Biochim. Biophys. Acta. 1398: 265-274.

Burd, C.G., and Dreyfuss, G. 1994. Conserved structures and diversity of functions of RNA-binding proteins. Science. 265: 615-621.

Burdon, R.H. 1987. Temperature and animal cell protein synthesis. Symp. Soc. Exp. Biol. 41: 113-133.

Busto, R., Dietrich, W.D., Globus, M.Y., Valdes, I., Scheinberg, P., and Ginsberg, M.D. 1987. Small differences in intraischemic brain temperature critically determine the extent of ischemic neuronal injury. Cereb. Blood Flow Metab. 7: 729-738.

Carpenter, C.D., Kreps, J.A., and Simon, A.E. 1994. Genes encoding glycine-rich Arabidopsis thaliana proteins with RNA-binding motifs are influenced by cold treatment and an endogenous circadian rhythm. Plant Physiol. 104: 1015-1025.

Chowdhury, A.K., and Steinberger, E. 1970. Early changes in the germinal epithelium of rat testes following exposure to heat. J. Reprod. Fertil. 22: 205-212.

Connolly, J.E., Boyd, R.J., and Calvin, J.W. 1962. The protective effect of hypothermia in cerebral ischemia: Experimental and clinical application by selective brain cooling in the human. Surgery. 52: 15-24.

Coop, A., Wiesmann, K.E., and Crabbe, M.J. 1998. Translocation of ß crystallin in neural cells in response to stress. FEBS Lett. 431: 319-321.

Cox, G., Moseley, P., and Hunninghake, G.W. 1993. Induction of heat-shock protein 70 in neutrophils during exposure to subphysiologic temperatures. J. Infect. Dis. 167: 769-771.

Cullen, K.E., and Sarge, K.D. 1997. Characterization of hypothermia-induced cellular stress response in mouse tissues. J. Biol. Chem. 272: 1742-1746.

Danno, S., Nishiyama, H., Higashitsuji, H., Yokoi, H., Xue J.H., Itoh, K., Matsuda, T., and Fujita, J. 1997. Increased transcript level of RBM3, a member of glycine-rich RNA-binding protein family, in human cells in response to cold stress. Biochem. Biophys. Res. Comm. 236: 804-807.

Danno, S., Itoh, K., Matsuda, T., and Fujita, J. 2000. Decreased expression of mouse Rbm^3, a cold-shock protein, in Sertoli cells of cryptorchid testis. Am. J. Pathol. In press.

Deman, J.J., and Bruyneel, E.A. 1977. Thermal transitions in the adhesiveness of HeLa cells: effects of cell growth, trypsin treatment and calcium. J. Cell Sci. 27: 167-181.

Derry, J.M., Kerns, J.A., and Francke, U. 1995. RBM3, a novel human gene in Xpl 1.23 with a putative RNA-binding domain. Hum. Mol. Genet. 4: 2307-2311.

Dreyfuss G, Matunis MJ, Pinol-Roma S, and Burd CG. 1993. hnRNP proteins and the biogenesis of mRNA. Annu. Rev. Biochem. 62: 289-321.

Dunn, M.A., Brown, K., Lightowlers, R., and Hughes, M.A. 1996. A low-temperature-responsive gene from barley encodes a protein with single-stranded nucleic acid-binding activity which is phosphorylated *in vitro*. Plant Mol. Biol. 30: 947-959.

El-Deiry, W.S., Tokino, T., Velculescu, V.E., Levy, D.B., Parsons, R., Trent, J.M., Lin, D., Mercer, W.E., Kinzler, K.W., and Vogelstein, B. 1993. WAF1, a potential mediator of p53 tumor suppression. Cell. 75: 817-825.

Easton, D.P., and Subjeck, J.R. 1997. The HSP110/SSE stress proteins- an overview. In: Guidebook to Molecular Chaperones and Protein-Folding Catalysts. M.-J. Gething, ed. Oxford Univ. Press, Oxford. p. 73-75.

Evans, R.G., Bagshaw, M.A., Gordon, L.F., Kurkjian, S.D., and Hahn, G.M. 1974. Modification of recovery from potentially lethal x-ray damage in plateau phase Chinese hamster cells. Radiat. Res. 59: 597-605.

Faraday, N., and Rosenfeld, B.A. 1998. *In vitro* hypothermia enhances platelet GPIIb-IIIa activation and P-selectin expression. Anesthesiology. 88: 1579-1585.

Frerichs, K.U., and Hallenbeck, J.M. 1998. Hibernation in ground squirrels induces state and species-specific tolerance to hypoxia and aglycemia: an *in vitro* study in hippocampal slices. J. Cereb. Blood Flow Metab. 18: 168-175.

Giard, D.J., Fleischaker, R.J., and Fabricant, M. 1982. Effect of temperature on the production of human fibroblast interferon. Proc. Soc. Exp. Biol. Med. 170: 155-169.

Glofcheski, D.J., Borrelli, M.J., Stafford, D.M., and Kruuv, J. 1993. Induction of tolerance to hypothermia and hyperthermia by a common mechanism in mammalian cells. J. Cell. Physiol. 156: 104-111.

Gon, Y., Hashimoto, S., Matsumoto, K., Nakayama, T., Takeshita, I., and Horie, T. 1998. Cooling and rewarming-induced IL-8 expression in human bronchial epithelial cells through p38 MAP kinase-dependent pathway. Biochem. Biophys. Res. Commun. 249: 156-160.

Gottlieb, T.M., and Oren, M. 1998. p53 and apoptosis. Semin. Cancer Biol. 8: 359-368.

Grand, R.J., Milner, A.E., Mustoe, T., Johnson, G.D., Owen, D., Grant, M.L., and Gregory, C.D. 1995. A novel protein expressed in mammalian cells undergoing apoptosis. Exp. Cell Res. 218: 439-451.

Gregory, C.D., and Milner, A.E. 1994. Regulation of cell survival in Burkitt lymphoma: implications from studies of apoptosis following cold-shock

treatment. Int. J. Cancer. 57: 419-426.
Guedez, L., Stetler-Stevenson, W.G., Wolff, L., Wang, J., Fukushima, P., Mansoor, A., and Stetler-Stevenson, M. 1998. In vitro suppression of programmed cell death of B cells by tissue inhibitor of metalloproteinases-1. J. Clin. Invest. 102: 2002-2010.
Hammond, E.M., Brunet, C.L., Johnson, G.D., Parkhill, J., Milner, A.E., Brady, G., Gregory, C.D., and Grand, R.J. 1998. Homology between a human apoptosis specific protein and the product of APG5, a gene involved in autophagy in yeast. FEBS Lett. 425: 391-395.
Hatayama, T., Tsujioka, K., Wakatsuki, T., Kitamura, T., and Imahara, H. 1992. Effects of low culture temperature on the induction of hsp70 mRNA and the accumulation of hsp70 and hsp105 in mouse FM3A cells. J. Biochem. 111: 484-490.
Hayashi, N., 1998. The control of brain tissue temperature and stimulation of dopamine-immune system to the severe brain injury patients. (In Japanese) Nippon Rinsho. 56: 1627-1635.
Heintzen, C., Melzer, S., Fischer, R., Kappeler, S., Apel, K., and Staiger, D. 1994. A light- and temperature-entrained circadian clock controls expression of transcripts encoding nuclear proteins with homology to RNA-binding proteins in meristematic tissue. Plant J. 5: 799-813.
Henle, K.J., and Leeper, D.B. 1979. Interaction of sublethal and potentially lethal 45 degrees-hyperthermia and radiation damage at 0, 20, 37 or 40 degrees C. Eur. J .Cancer. 15: 1387-1394.
Hochachka, P.W. 1986. Defense strategies against hypoxia and hypothermia. Science. 231: 234-241.
Holahan, E.V., Bushnell, K.M., Highfield, D.P., and Dewey, W.C. 1982. The effect of cold storage of mitotic cells on hyperthermic killing and hyperthermic radiosensitization during G1 and S. Radiat. Res. 92: 568-573.
Holland, D.B., Roberts, S.G., Wood, E.J., and Cunliffe, W.J. 1993. Cold shock induces the synthesis of stress proteins in human keratinocytes. J. Invest. Dermatol. 101: 196-199.
Hughes, M.A., Dunn, M.A., and Brown, A.P.C. 1999. Approaches to the analysis of cold-induced barley genes isolated through differential screening of a cDNA library. In: Environmental Stress and Gene Regulation. K.B. Storey, ed. Bios Scientific Pub. Ltd., Oxford. p.139-158.
Johanson, K.J., Wlodek, D., and Szumiel, I. 1983. Effect of ionizing radiation and low temperature on L5178Y-R and L5178Y-S cells. II. Cell survival and DNA strand breaks after roentgen or gamma irradiation. Acta Radiol. Oncol. 22: 71-76.
Jones, P.G. And Inouye, M. 1994. The cold-shock response-a hot topic. Mol. Micro. Biol. 11: 811-818.
Kaneko, Y., Kimura, T., Kishishita, M., Noda, Y., and Fujita, J. 1997a. Cloning of apg-2 encoding a novel member of heat shock protein 110 family. Gene. 189: 19-24.
Kaneko, Y., Kimura, T., Nishiyama, H., Noda, Y., and Fujita, J. 1997b. Developmentally regulated expression of APG-1, a member of heat shock protein 110 family in murine male germ cells. Biochem. Biophys. Res.

Commun. 233: 113-116.

Kaneko, Y., Nishiyama, H., Nonoguchi, K., Higashitsuji, H., Kishishita, M., and Fujita, J. 1997c. A novel hsp110-related gene, apg1, that is abundantly expressed in the testis responds to a low temperature heat shock rather than the traditional elevated temperatures. J. Biol. Chem. 272: 2640-2645.

King, L, and Weber, G. 1986. Conformational drift and cryoinactivation of lactate dehydrogenase. Biochem. 25: 3637-3640.

Karsch-Mizrachi, I., and Haynes, S.R. 1993. The Rb97D gene encodes a potential RNA-bindingprotein required for spermatogenesis in Drosophila. Nucleic Acids Res. 21: 2229-2235.

Kojima, R., Randall, J., Brenner, B.M., and Gullans, S.R. 1996. Osmotic stress protein 94 (Osp94). A new member of the Hsp110/SSE gene subfamily. J. Biol. Chem. 271: 12327-12332.

Kruman, I.I., Gukovskaya, A.S., Petrunyaka, V.V., Beletsky, I.P., and Trepakova, E.S. 1992. Apoptosis of murine BW 5147 thymoma cells induced by cold shock. J. Cell. Physiol. 153: 112-117.

Laios, E., Rebeyka, I.M., and Prody, C.A. 1997. Characterization of cold-induced heat shock protein expression in neonatal rat cardiomyocytes. Mol. Cell. Biochem. 173: 153-159.

Li, G.C., and Hahn, G.M. 1980. Adaptation to different growth temperatures modifies some mammalian cell survival responses. Exp. Cell Res. 128: 475-479.

Liepins, A., and Younghusband, H.B. 1985. Low temperature-induced cell surface membrane vesicle shedding is associated with DNA fragmentation. Exp. Cell Res. 161: 525-532.

Lindquist, S., and Craig, E.A. 1988. The heat-shock proteins. Ann. Rev. Genet. 22: 631-677.

Liu, A.Y., Bian, H., Huang, L.E., and Lee, Y.K. 1994. Transient cold shock induces the heat shock response upon recovery at 37 degrees C in human cells. J. Biol. Chem. 269: 14768-14775.

Liu, Z.G., Baskaran R., Lea-Chou, E.T., Wood, L.D., Chen, Y., Karin, M., and Wang, J.Y. 1996. Three distinct signalling responses by murine fibroblasts to genotoxic stress. Nature. 384: 273-276.

Luking, A., Stahl, U., and Schmidt, U. 1998. The protein family of RNA helicases. Crit. Rev. Biochem. Mol. Biol. 33: 259-296.

Ma, K., Inglis, J.D., Sharkey, A., Bickmore, W.A., Hill, R.E., Prosser, E.J., Speed, R.M., Thomson, E.J., Jobling, M., Taylor, K., Wolfe, J., Cooke, H.J., Hargreave, T.B., and Chandley, A.C. 1993. A Y chromosome gene family with RNA-binding protein homology: candidates for the azoospermia factor AZF controlling human spermatogenesis. Cell. 75: 1287-1295.

Marino, F., and Booth, J. 1998. Whole body cooling by immersion in water at moderate temperatures. J. Sci. Med. Sport. 1: 73-82.

Marion, D.W., Penrod, L.E., Kelsey, S.F., Obrist, W.D., Kochanek, P.M., Palmer, A.M., Wisniewski, S.R., and DeKosky, S.T. 1997. Treatment of traumatic brain injury with moderate hypothermia. N. Engl. J. Med. 336: 540-546.

Matz, J.M., Blake, M.J., Tatelman, H.M., Lavoi, K.P., and Holbrook, N.J. 1995.

Characterization and regulation of cold-induced heat shock protein expression in mouse brown adipose tissue. Am. J. Physiol. 269: R38-47.

Mauney, M.C., and Kron, I.L. 1995. The physiologic basis of warm cardioplegia. Ann. Thorac. Surg. 60: 819-823.

Meijer, J.H., and Rietveld, W.J. 1989. Neurophysiology of the suprachiasmatic circadian pacemaker in rodents. Physiol. Rev. 69: 671-707.

Morimoto, R. I., Tissieres, A., and Georgopoulos, C. 1994. Progress and perspectives on the biology of heat shock proteins and molecular chaperones. In: The Biology of Heat Shock Proteins and Molecular Chaperones. R.I. Morimoto, A. Tissieres, A., and Georgopoulos, C., eds. Cold Spring Harbor Lab. Press, New York. p.1-30.

Negrutskii, B.S., Stapulionis, R., and Deutscher, M.P. 1994. Supramolecular organization of the mammalian translation system. Proc. Natl. Acad. Sci. USA. 91: 964-968.

Nelson, R.J., Kruuv, J., Koch, C.J., and Frey, H.E. 1971. Effect of sub-optimal temperatures on survival of mammalian cells. Exp. Cell Res. 68: 247-252.

Ning, X.H., Xu, C.S., Song, Y.C., Xiao, Y., Hu, Y.J., Lupinetti, F.M., and Portman, M.A. 1998. Hypothermia preserves function and signaling for mitochondrial biogenesis during subsequent ischemia. Am. J. Physiol. 274: H786-793.

Nishiyama, H., Danno, S., Kaneko, Y., Itoh, K., Yokoi, H., Fukumoto, M., Okuno, H., Millan, J.L., Matsuda, T., Yoshida, O., and Fujita, J. 1998a. Decreased expression of cold-inducible RNA-binding protein (CIRP) in male germ cells at elevated temperature. Am. J. Pathol. 152: 289-296.

Nishiyama, H., Higashitsuji, H., Yokoi, H., Itoh, K., Danno, S., Matsuda, T., and Fujita, J. 1997a. Cloning and characterization of human CIRP (cold-inducible RNA-binding protein) cDNA and chromosomal assignment of the gene. Gene. 204: 115-120.

Nishiyama, H., Itoh, K., Kaneko, Y., Kishishita, M., Yoshida, O., and Fujita, J. 1997b. A glycine-rich RNA-binding protein mediating cold-inducible suppression of mammalian cell growth. J. Cell Biol. 137: 899-908.

Nishiyama, H., Xue, J.H., Sato, T., Fukuyama, H., Mizuno, N., Houtani, T., Sugimoto, T., and Fujita, J. 1998b. Diurnal change of the cold-inducible RNA-binding protein (Cirp) expression in mouse brain. Biochem. Biophys. Res. Commun. 245: 534-538.

Nomura, N., Nagase, T., Miyajima, N., Sazuka, T., Tanaka, A., Sato, S., Seki, N., Kawarabayasi, Y., Ishikawa, K., and Tabata, S. 1994. Prediction of the coding sequences of unidentified human genes. II. The coding sequences of 40 new genes (KIAA0041-KIAA0080) deduced by analysis of cDNA clones from human cell line KG-1. DNA Res. 1: 223-229.

Oh, H.J., Easton, D., Murawski, M., Kaneko, Y., and Subjeck, J.R. 1999. The chaperoning activity if Hsp110: Identification of functional domains by use of targeted deletions. J. Biol. Chem. 274: 15712-15718.

Ohnishi, T., Wang, X., Ohnishi, K., and Takahashi, A. 1998. p53-dependent induction of WAFi by cold shock in human glioblastoma cells. Oncogene. 16: 1507-1511.

Ornelles, D.A., Fey, E.G., and Penman, S. 1986. Cytochalasin releases mRNA from the cytoskeletal framework and inhibits protein synthesis. Mol. Cell. Biol. 6: 1650-1662.

Ota, T., Hanada, K., and Hashimoto, I. 1996. The effect of cold stress on UVB injury in mouse skin and cultured keratinocytes. Photochem. Photobiol. 64: 984-987.

Pain, R.H. 1987. Temperature and macromolecular structure and function. Symp. Soc. Exp. Biol. 41: 21-33.

Perotti, M., Toddei, F., Mirabelli, F., Vairetti, M., Bellomo, G., McConkey, D.J., and Orrenius, S. 1990. Calcium-dependent DNA fragmentation in human synovial cells exposed to cold shock. FEBS Lett. 259: 331-334.

Phillips, R.A., and Tolmach, L.J. 1966. Repair of potentially lethal damage in x-irradiated HeLa cells. Radiat Res. 29: 413-432.

Porter, K.R., and Tucker, J.B. 1981. The ground substance of the living cell. Sci. Am. 244: 56-67.

Portman, D.S., and Dreyfuss, G. 1994. RNA annealing activities in HeLa nuclei. EMBO J. 13: 213-221.

Puigserver, P., Wu, Z., Park, C.W., Graves, R., Wright, M., and Spiegelman, B.M. 1998. A cold-inducible coactivator of nuclear receptors linked to adaptive thermogenesis. Cell. 92: 829-839.

Rao, P.N., and Engelberg, J. 1965. HeLa cells: effects of temperature on the life cycle. 148: 1092-1039.

Reijo, R., Lee, T.Y., Salo, P., Alagappan, R., Brown, L.G., Rosenberg, M., Rozen, S., Jaffe, T., Straus, D., Hovatta, O., *et al.* 1995. Diverse spermatogenic defects in humans caused by Y chromosome deletions encompassing a novel RNA-binding protein gene. Nat. Genet. 10: 383-393.

Robinow, S., Campos, A.R., Yao, K.M., and White, K. 1988. The elav gene product of Drosophila, required in neurons, has three RNP consensus motifs. Science. 242: 1570-1572.

Rule, G.S., Law, P., Kruuv, J., and Lepock, J.R. 1980. Water permeability of mammalian cells as a function of temperature in the presence of dimethylsulfoxide: correlation with the state of the membrane lipids. J. Cell. Physiol. 103: 407-416.

Russotti, G., Brieva, T.A., Toner, M., and Yarmush, M.L. 1996. Induction of tolerance to hypothermia by previous heat shock using human fibroblasts in culture. Cryobiol. 33: 567-580.

Sadaie, M.R., Tschachler, E., Valerie, K., Rosenberg, M., Felber, B.K., Pavlakis, G.N., Klotman, M.E., and Wong-Staal, F. 1990. Activation of tat-defective human immunodeficiency virus by ultraviolet light. New Biol. 2: 479-486.

Sarge, K.D. 1995. Male germ cell-specific alteration in temperature set point of the cellular stress response. J. Biol. Chem. 270: 18745-18748.

Setchell, B.P. 1982. Spermatogenesis and spermatozoa. In: Reproduction in Mammals:1 Germ Cells and Fertilization. C.R. Austin, and R.V. Short, eds. Cambridge Univ. Press, New York. p. 63-101.

Setlow, V.P., Roth, S., and Edidin, M. 1979. Effects of temperature on glycosyltransferase activity in the plasma membrane of L cells. Exp. Cell

Res. 121: 55-61.

Shapiro, I.M., and Lubennikova, E.I. 1968. Population kinetics of cells in tissue culture incubated at low temperature. Exp. Cell Res. 49: 305-316.

Sheikh, M.S., Carrier, F., Papathanasiou, M.A., Hollander, M.C., Zhan, Q., Yu, K., and Fornace, A.J. Jr. 1997. Identification of several human homologs of hamsterDNA damage-inducible transcripts. Cloning and characterization of a novel UV-inducible cDNA that codes for a putative RNA-Binding binding protein. J. Biol. Chem. 272: 26720-26726.

Shinozaki, K., and Yamaguchi-Shinozaki, K. 1997. Gene expression and signal transduction in water-stress response. Plant Physiol. 115: 327-334.

Sisken, J.E., Morasca, L., and Kibby, S. 1965. Effects of temperature on the kinetics of the mitotic cycle of mammalian cells in culture. Exp. Cell Res. 39: 103-116.

Shodell, M. 1975. Reversible arrest of mouse 3T6 cells in G2 phase of growth by manipulation of a membrane-mediated G2 function. Nature. 256: 578-580.

Soloff, B.L., Nagle, W.A., Moss, A.J. Jr., and Henle, K.J., Crawford, JT. 1987. Apoptosis induced by cold shock *in vitro* is dependent on cell growth phase. Biochem. Biophys. Res. Commun. 145: 876-883.

Stapulionis, R., Kolli, S., and Deutscher, M.P. 1997. Efficient mammalian protein synthesis requires an intact F-actin system. J. Biol. Chem. 272: 24980-24986.

Thieringer, H.A., Jones, P.G., and Inouye, M. 1998. Cold shock and adaptation. Bioessays. 20: 49-57.

Tissieres, A., Mitchell, H.K., and Tracy, U.M. 1974. Protein synthesis in salivary glands of Drosophila melanogaster: relation to chromosome puffs. J. Mol. Biol. 84: 389-398.

Ueno, A.M., Goldin, E.M., Cox, A.B., and Lett, J.T. 1979. Deficient repair and degradation of DNA in X- Irradiated irradiated L5178Y S/S cells: cell-cycle and temperature dependence. Radiat. Res. 79: 377-389.

Uochi, T., and Asashima, M. 1998. XCIRP (Xenopus homolog of cold-inducible RNA-binding protein) is expressed transiently in developing pronephros and neural tissue. Gene 211: 245-250.

van Nocker, S., and Vierstra, R.D. 1993. Two cDNAs from Arabidopsis thaliana encode putative RNA binding proteins containing glycine-rich domains. Plant Mol. Biol. 21: 695-699.

van Rijn, J., van den Berg, J., Kipp, J.B., Schamhart, D.H., and van Wijk, R. 1985. Effect of hypothermia on cell kinetics and response to hyperthermia and X rays. Radiat. Res. 101: 292-305.

Vigh, L., Maresca, B., and Harwood, J.L. 1998. Does the membrane's physical state control the expression of heat shock and other genes? Trends Biochem. Sci. 23: 369-374.

Wassmann, H., Greiner, C., Hulsmann, S., Moskopp, D., Speckmann, E.J., Meyer, J., and van Aken, H. 1998. Hypothermia as cerebroprotective measure. Experimental hypoxic exposure of brain slices and clinical application in critically reduced cerebral perfusion pressure. Neurol. Res. 20: S61-S65.

Watanabe, I., and Okada, S. 1967. Effects of temperature on growth rate of cultured mammalian cells (L5178Y). J. Cell. Biol. 32: 309-323.

Watson, P.F., and Morris, G.J. 1987. Cold shock injury in animal cells. Symp. Soc. Exp. Biol. 41: 311-340.

Weisenberg, R.C. 1972. Microtubule formation in vitro in solutions containing low calcium concentrations. Science. 177: 1104-1105. Yasuda, K., Nakai, A., Hatayama, T., and Nagata, K. 1995. Cloning and expression of murine high molecular mass heat shock proteins, HSP105. J. Biol. Chem. 270: 29718-29723.

Xue, J.H., Fukuyama, H., Nonoguchi, K., Kaneko, Y., Kido, T., Fukumoto, M., Fujibayashi, Y., Itoh, K, and Fujita, J. 1998. Induction of Apg-1, a member of the heat shock protein110 family, following transient forebrain ischemia in the rat brain. Biochem. Biophys. Res. Commun. 247: 796-801.

Xue, J.H., Nonoguchi, K., Fukumoto, M., Sato, T., Nishiyama, H., Higashitsuji, H., Itoh, K., and Fujita, J. 1999. Effects of ischemia and H_2O_2 on the cold stress protein cirp expression in rat neuronal cells. Free Rad. Biol. Med. 27: 1238-1244.

Index

A

Abscisic acid 94
Acclimation, *see* Cold acclimation
Acyl-lipid desaturases 63, *see also* Enzymes
Adaptation 41, *see also* Cold shock, Cold adaptation, Cold stress, Cold acclimation
  of translation factory 8
  physiology 43
Adaptive thermogenesis 113
Anabaena 64
Antarctica 61
Antifreeze proteins 89
APG 113, 132
ASP 132

B

Bacillus subtilis 27
Biotechnology 113
Brain damage, treatment of 113

C

CAPs, *see* Cold acclimation proteins
Cell proliferation 113
Chaperones 34, 61, 113
Chromosome, *see also* DNA
  adjustment 28
  dynamics 7
CIPs, *see* Cold induced proteins
CIRP 113, 124-125, 127-129, 132, *see also* RNA binding protein
  function 129
  induction 125, 127, 128
  structure 125
clp genes 64
Clp 61, 71, *see also* Cold induced proteins
Cold acclimation 27, 30, 41, 46, 49, 61, 86
  in plants 86
  in psychrotrophic bacteria 41, 46
  proteins 27, 30, 41, 49
Cold adaptation, *see also* Cold acclimation
  enzymes 44
Cold induced genes 61, 63
  common features 73
  expression 61
  regulation 63
  regulatory regions 73
Cold induced proteins 27, 30, 41, 47-51, 61
  expression 50
  kinetics of expression 50
  synthesis of 41, 47
Cold inducible RNA binding protein 113, 124, 132, *see also* CIRP
Cold regulated gene expresion 100
Cold sensing mechanism 61
Cold shock, *see also* Cold adaptation, Cold stress, Cold acclimation
  acclimatization 27
  adaptation of translation 8
  cytosolic response 29
  domain 16, 91, 92
  domain proteins 91
  effect on growth 46
  enzymes 44, *see also* Enzymes
  evolution 16
  in *Bacillus subtilis* 27
  in bacteria 36
  in cyanobacteria 61
  in *E. coli* 5
  in mammalian cells 113
  in psychrotrophic bacteria 41, 46
  in plants 86
  induction 32, *see also* Cold adaptation, Cold stress, Cold acclimation
  membrane changes 6
  protein changes 8
  protein folding 32

protein in mammals 124, *see also* CIRP
protein synthesis 41, 45, 47
proteins 13, 14, 33-35, 41, 48, 49, 50, 51, 88
proteins, as RNA chaperones 34, 61, 113
proteins, function of 33, 34
proteins of *Bacillus subtilis* 27
proteins of *E. coli* 7
regulation of genes 61
stress induction 32
translation adaptation 8
Cold shock proteins 13, 14, 33-35, 41, 48, 49, 50, 51, 88
Cold shock response, *see also* Cold shock
in *Bacillus subtilis* 27
in bacteria 36
in cyanobacteria 61
in mammalian cells 113
in plants 86
in psychrotrophic bacteria 41, 46
of membrane lipid composition 6
Cold stress 27, *see also* Cold shock, Cold adaptation, Cold acclimation
induction 32
mild 122
proteins 88
severe 120
Cold tolerance, in mammals 113
crh genes 64, *see also* Genes
csp genes 10, *see also* Genes
Csp 49, *see also* Cold shock proteins
expression 52
cspA, *see also* Genes
expression 11
regulation of expression 11
CspA 13, 41, 50, 51, *see also* Proteins
family of *E. coli* 14
structure and function 13
CspB 35, *see also* Proteins
from *Bacillus subtilis* 35
structure of 35

X-ray crystallography 35
Csps 50, 51, *see also* Cold shock proteins
Cyanobacteria 61, 63
Cytosolic protein composition 27
Cytosolic protein synthesis 27
Cytosolic response 29

D

des genes 64, 74, *see also* Genes
Desaturases 63, *see also* Enzymes
DNA 27, *see also* Chromosome topology 27
Drought 95, 100

E

E. coli 5
Energy generation 7
Environmental temperature 1, 73, 86
Enzymes, *see also* Proteins
acyl-lipid desaturases 63
cold adapted 44
cold shock enzymes 44
desaturases 63
fatty acid desaturase 61, 63, 67
functions of 67
histidine kinase 76
MAP kinase cascade 95
prolyl isomerase 32
RNA helicase 61, 70
Evolution of cold shock domain 16
Expression of Csp 52, *see also* Cold shock proteins

F

Fatty acid desaturase 61, 63, 67
functions of 67
Freezing 95
growth at 41
stress 92, *see also* Cold shock, Cold stress

G

Gene expression 11, 73, 86
   cold regulated 100
   in mammals 120
   in plants 95
   induced by cold stress 120
Genes
   APG 132
   ASP 132
   clp genes 64
   cold inducible 132, 133
   crh genes 64
   csp genes 10
   cspA, expression 11
   cspA, regulation of expression 11
   des genes 64, 74
   HSP 132
   IL-8 132
   KIAA0058 132
   lti2 64, 72
   RBM 132
   regulation of *cspA* expression 11
   rpb genes 64
   rpsU 64
   WAF1 132
Genetic modification 100
Growth
   effect of cold shock on 46
   kinetics at low temperature 43

H

Helicase, *see* RNA helicase
Histidine kinase 76
HSP 132
   transcripts 123
HSP110 family 113

I

IL-8 132

K

KIAA0058 132

L

Lag phase 51
Lipid composition 43, 113
Lipids 61, *see also* Membranes
Low temperature
   growth kinetics 43
   lipid composition 43
   protein changes 8
   protein content 45
   protein folding 32
   protein synthesis 45
lti2 64, 72

M

Mammalian cells 113
Mammals 113
MAP kinase cascade 95
Medical science 113
Membranes 65, 113
   adaptation 28
   composition 27
   fluidity 41, 43, 61, 73, *see also* Fatty acid desaturase
   lipids 113
Mesophilic bacteria 29
Metabolic activity 41
Mild cold stress 122
Molecular responses 86

O

Osmotic stress 92, 95

P

Photosynthesis 87
Phylogenetic relationships 122
Physiological response 113
Physiology

of cold adapted microorganisms 43
regulation 51
Plants 86
Prolyl isomerase 32, *see also* Enzymes
Proteins, *see also* Cold shock proteins, Enzymes, Cold induced proteins, Cold acclimation
  acyl-lipid desaturases 63
  antifreeze 89
  APG 113, 132
  as RNA chaperones 34, 61, 113
  ASP 132
  CAPs 49, *see also* Cold acclimation proteins
  CIPs, *see* Cold induced proteins
  CIRP 113, 124, 127-129, 132
  CIRP function 129
  CIRP induction 125, 127, 128
  CIRP structure 125
  Clp 61, 71, *see also* Cold induced proteins
  cold acclimation 27, 30, 41
  cold induced 27, 30, 41, 48, 49, 50, 61
  cold induced, expression 50
  cold induced, kinetics of expression 50
  cold induced, synthesis of 41, 47
  cold inducible RNA binding, *see* CIRP
  cold shock domain 91
  cold shock proteins 33, 34, 38, 41, 48, 49
  cold shock proteins, as RNA chaperones 34, 61, 113
  cold shock proteins, function of 33, 34
  cold shock proteins in mammals 124, *see also* CIRP
  cold shock proteins of *Bacillus subtilis* 27
  cold shock proteins, synthesis of 41, 45, 47
  cold stress in plants 88
  content at low temperature 45
  Csp 13, 14, 33-35, 41, 48, 49, 50, 51, 88, *see also* Cold shock proteins
  Csp expression 52
  CspA 41, 50, 51
  CspA family of *E. coli* 14
  CspA structure and function 13
  CspB, from *Bacillus subtilis* 35
  CspB, structure of 35
  CspB, X-ray crystallography 35
  Csps 13, 14, 33-35, 41, 48, 49, 50, 51, 88
  cytosolic protein composition 27
  cytosolic protein synthesis 27
  desaturases 63
  expression 52
  expression of Csp 52, *see also* Cold shock proteins
  fatty acid desaturase 61, 63, 67
  fatty acid desaturase, functions of 67
  folding, 32, 61
  folding at low temperature, 32
  folding, prolyl isomerase 32
  helicase *see* RNA helicase
  histidine kinase 76
  HSP 132
  HSP family 113
  HSP transcripts 123
  IL-8 132
  induced by cold stress 27
  KIAA0058 132
  kinetics of expression 50
  low temperature protein content 45
  low temperature protein folding 32
  low temperature protein synthesis 45
  MAP kinase cascade 95
  medical science 113
  molecular changes at low temperature 8
  production 113
  prolyl isomerase 32
  RBM 132

Rbp family 68
RNA helicase 61, 70
RNA chaperone 34, 61, 113
RNA binding proteins 61, 68, 113, 127-129, see also CIRP
Rpb family 68
S21 70
stress induction 32
structure and function of CspA 13
synthesis 41
synthesis at low temperature 45, 47
synthesis, effect of cold shock 47
synthesis in mammalian cells 113
synthesis, repression of 41
translation 61
WAF1 132
Psychrotrophic bacteria 41, 46
Psychrotrophs 41, 46

R

RBM 132
Recovery from stress 118
Regulation of *cspA* expression 11
Respiration 87
Ribosome 70
RNA 34, 61, see also DNA, Chromosome
  binding protein 61, 113, see also CIRP
  chaperone 61, 68, 113, 127-129
  helicase 34, 61, 70
rpb genes 64
Rpb family 68
rpsU 64

S

Salinity 95
Severe cold stress 120
Signal perception 86, 95
Signal transduction pathways 95

Storage of mammalian organs 113
Stress tolerance 95
  in plants 86
  inducible 86
Stress induction 32, see also Cold shock, Cold adaptation, Cold stress
Stress recovery 118
Stress response
  in mammals 118, 119
  modification of 118
Structure and function of CspA 13
Synechococcus 64
Synechocystis 63, 64

T

Temperature 1, 73, 86
  adaptation 29, see also Cold shock, Cold adaptation, Cold stress
  change 1
  stress tolerance 100, see also Cold shock, Cold stress
Transcriptional activation 100
Transcriptional activation factors 100
Transcriptional control 86
Transduction pathways 86
Translation 61

W

WAF1 132

Books of Related Interest

Gene Cloning and Analysis: Current Innovations Brian C. Schaefer (Ed.)	1997
An Introduction to Molecular Biology Robert C. Tait	1997
Genetic Engineering with PCR Robert M. Horton and Robert C. Tait (Eds.)	1998
Prions: Molecular and Cellular Biology David A. Harris (Ed.)	1999
Probiotics: A Critical Review Gerald W. Tannock (Ed.)	1999
Peptide Nucleic Acids: Protocols and Applications Peter E. Nielsen and Michael Egholm (Eds.)	1999
Intracellular Ribozyme Applications: Principles and Protocols John J. Rossi and Larry Couture (Eds.)	1999
NMR in Microbiology: Theory and Applications Jean-Noël Barbotin and Jean-Charles Portais (Eds.)	2000
Molecular Marine Microbiology Douglas H. Bartlett (Ed.)	2000
Oral Bacterial Ecology: The Molecular Basis Howard K. Kuramitsu and Richard P. Ellen (Eds.)	2000
Prokaryotic Nitrogen Fixation: A Model System Eric W. Triplett (Ed.)	2000

For further information on these books contact:

Horizon Scientific Press, P.O. Box 1, Wymondham, Norfolk, NR18 0EH, U.K.
Tel: +44(0)1953-601106. Fax: +44(0)1953-603068. Email: mail@horizonpress.com

Our Web site has details of all our books including full chapter abstracts, book reviews, and ordering information:

www.horizonpress.com